T0280642

Textile Science and Clothing Technology

Series editor

Subramanian Senthilkannan Muthu, Kowloon, Hong Kong

More information about this series at http://www.springer.com/series/13111

Subramanian Senthilkannan Muthu
Editor

Green Composites

Sustainable Raw Materials

 Springer

Editor
Subramanian Senthilkannan Muthu
SgT Group and API
Hong Kong, Kowloon
Hong Kong

ISSN 2197-9863   ISSN 2197-9871   (electronic)
Textile Science and Clothing Technology
ISBN 978-981-13-4711-5   ISBN 978-981-13-1969-3   (eBook)
https://doi.org/10.1007/978-981-13-1969-3

This book is dedicated to:
The lotus feet of my beloved
Lord Pazhaniandavar
My beloved late Father
My beloved Mother
My beloved Wife Karpagam and
Daughters—Anu and Karthika
My beloved Brother
Last but not least
To everyone working in the textile
sector to make it SUSTAINABLE

Contents

Production of Green Composites from Various Sustainable Raw Materials

Akshay C. Jadhav, Pintu Pandit, T. Nadathur Gayatri,
Pravin P. Chavan and Nilesh C. Jadhav

Abstract The developing anxiety towards prevention of ecological destruction and the unfilled requirement for more adaptable environmental friendly materials has prompted expanding interest about polymer composites, originating from sustainable sources and biodegradable plant materials, particularly from forests. The composites for the most part are referred to as 'green' and can be used in industrial applications. Green composites do not harm the environment much and could be satisfactory alternatives to petroleum-based polymers and polymer composites. Using renewable resources like vegetable oils, carbohydrates and proteins, to develop biopolymer matrices like in cellulose-reinforced green composites, it is possible to minimize the consumption of fossil oil resources. Vegetable oils are not costly, easily sourced and could be used to synthesize sustainable polymers. These have widened the utilization of plant fibres as reinforcements and have increased the possibility for sustainable and 'biodegradable' composites, which can be called 'green' composites as they satisfy the criteria of 'green materials'. Thus, the challenge to obtain 'green' composite involves obtaining 'green' polymers functioning as matrices in the production of composite materials. This chapter considers the materials and methods utilized for the fabrication and particularly the utilization of green composites in different technological fields. Furthermore, a discussion on the sustainability of major raw materials utilized in green composites is provided in this chapter.

Keywords Green composite · Biodegradable · Natural fibres · Sustainability
Green materials

Natural fibres have benefits like less density, good stiffness, reasonable mechanical properties, high disposability and renewability (Kim et al. 2002; Reddy and Yang 2005a). Fibre surface treatments like mercerization, isocyanate treatment, acetyla-

A. C. Jadhav · P. Pandit (✉) · T. N. Gayatri · P. P. Chavan · N. C. Jadhav
Department of Fibres and Textile Processing Technology, Institute of Chemical Technology,
University Under Section-3 of UGC Act 1956, N. P. Marg, Matunga, Mumbai 400019, India
e-mail: pintupanditict@gmail.com

© Springer Nature Singapore Pte Ltd. 2019
S. S. Muthu (ed.), *Green Composites*, Textile Science and Clothing Technology,
https://doi.org/10.1007/978-981-13-1969-3_1

tion and hydrogen peroxide treatment are applicable processes in gaining improvement of properties (Reddy and Yang 2005a, b; Huda et al. 2007). Most acceptable surface modification of natural fibres is obtained with sodium hydroxide (NaOH) treatment. Major improvement in the mechanical property of cellulose-based natural fibres is often obtained by alkali treatment (Mohanty et al. 2000). Natural fibre-based cellulose is primarily gaining attention as their application in advanced engineering materials is increasing. It has been estimated that there are a minimum of one thousand forms of plants that produce usable fibres. Fibres are mostly extracted from different parts of plants, for example, cotton and kapok from seed; sisal and pineapple from the leaf; jute, flax, kenaf, ramie and hemp are obtained from the stem; fibre and monocot genus from the fruit (Fig. 1).

Today, consumers are showing higher levels of environmental awareness, and seeking eco-friendly products, and so untapped fibre resources ought to be more explored for their potential application in textile. Attempts are being created to use annually renewable lignocellulosic agricultural by-product, like rice husks, corn husks, corn stalks and pineapple leaves, as an alternate supply for natural fibres. Nearly one hundred million heaps of fibres are consumed annually within the world. Benefits of natural fibres over synthetic fibres embrace low price, less density, recyclability and biodegradability and their carbon dioxide neutral life cycle (Baley 2002). The plant fibres are composed of cellulose, hemicelluloses and lignin. Surface impurities such as pectins and wax substances are also present. The foremost vital chemical component of plant fibres is cellulose. Cellulose contains three hydroxyl groups that within organic compound cellulose structure forms hydrogen bonds, whereas the third forms intermolecular hydrogen bonds. Lignin is an

Fig. 1 Natural fibres used in green composites

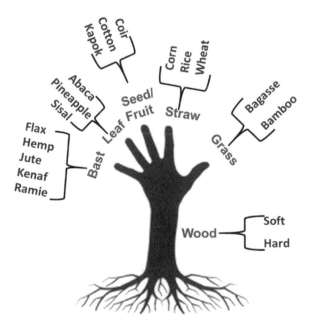

advanced three-dimensional chemical compound whose function is to attach elementary fibres in bundles; thus, additionally lignin content makes the fibre structure stiffer. Hemicellulose consists chiefly of xylan, polyuronide and hexosan. The chemical composition of the natural fibres relies on the sort, origin, age of the fibre and processes undergone during the extraction methodology.

There are at least one thousand forms of plants which bear useful fibres (Van Voorn et al. 2001). Natural fibres that are obtained from plant sources are mostly found in tropical areas and are out there throughout the terrestrial forests. Nowadays, these plant fibres are considered eco-friendly as they possess biodegradability and have renewable properties (Das and Chakraborty 2008; John and Anandjiwala 2008). Natural fibres are applied in the fields of textile industry, paper producing industry and bioenergy industries due to their overall properties. Natural plant fibres are usually classified and supported by their origin (e.g. plant source, animal source or mineral source). These plant or vegetable fibres are further categorized into subgroups consistent with their supply [e.g. stem (bast) fibres, fruit or seed fibres, leaf fibres] (Mwaikambo and Ansell 2002; Bledzki and Gassan 1999). Natural fibres are obtained from completely different components of plants, and these raw materials are extensively used for making green composites (Robson 1993).

1 Properties of Sustainable Raw Materials as Reinforcing Materials

Bast fibres are normally extracted from the stem of the plants. These fibres are the conductive cells of phloem and provide strength to the stem of the plant. Some of the important bast fibres, available in the market, are flax, hemp, ramie, jute, kenaf, etc. Recently, some of the bast fibres also have been extracted from the stem of the wild plants like nettle, wisteria, linden. Normally, bast fibres have high tensile strength and low extensibility and are used for making ropes, yarn, paper and composite materials (Kiruthika 2017).

The chemical composition of a fibre affects its appearance, structure, properties and processability. In this context, compositions and the properties of the extracted bast fibres have been reported (Table 1). Hemicellulose content of the jute fibre is quite higher rather than the flax and the hemp fibre. Lignin, the binding cementing constituent that holds the individual cells together, is present in higher amount in the jute fibre (11.8–14.2%), followed by the hemp and flax fibre. This might be possibly due to the presence of more amorphous hemicellulose (40%) in it that is significantly higher than flax and hemp fibres. Due to the occurrence of these less crystalline regions, jute fibre is more accessible to water, dyes and other finishing chemicals. Indeed, flax and hemp also have high moisture regain of 13%, which is more than the jute fibre (Chand and Hashmi 1993; Das et al. 2015; Teli et al. 2018).

Table 1 Physical and chemical properties of important bast fibres (Kiruthika 2017; Chand and Hashmi 1993; Das et al. 2015)

Composition and properties	Jute	Flax	Hemp
Cellulose content (%)	59–72	65–87	Under 80%
Hemicellulose content (%)	12–15	Nil	Nil
Lignin (%)	11.8–14.2	Small	2–3
Crystallinity index	50–55	65–70	66.9
Moisture regain (%)	13.8	12	13
Strength (g/tex)		30–60	23–50

Hemp fibre is also similar to flax fibre. However, the amount of the lignin present in the hemp fibre is slightly more than the extent of the lignin present in the flax fibre. Ramie fibre is an example of the one of the strongest bast fibres, and like cotton fibre, its strength has been increased in the wet condition. However, ramie fibre lacks resiliency, elasticity property and it has very low abrasion resistance (Teli and Pandit 2017a; Basak et al. 2016; Ammayappan et al. 2016).

Flax fibre also contains lignin but in lesser amount compared to the jute fibre, and the initial modulus of flax is even more than jute. Flax fibre is another example of the bast fibre and mostly used by the composite industries. Normally, the flax fibre (light greyish in colour) has been extracted from the stem of the flax plant, and by nature, it is soft, lustrous, flexible and strongest fibre. It contains 65–68% cellulose with small amount of the lignin present in its structure. Spiral angle of the secondary cell wall of the cellulose of the flax fibre is in the range of 6–8°. The tenacity range of the flax fibre is around 6.5–8 gm/den, while the length of the fibre is in the range of 18–30 inches. Specific gravity of the flax fibre is around 1.54 g/cc (Ammayappan et al. 2016; Das 2017).

Jute fibre is a multi-cellular lignocellulosic fibre with the initial modulus of 1150 g/tex, and the tenacity is in the rage of 27–44 g/tex. Cellulose part of the jute fibre is composed of linear chain of the glucopyranose unit connected by 1,4-β-glucosidic linkage. It is crystalline in nature. Spiral angle in the secondary cell wall of the cellulose is around 6–8° (Ammayappan et al. 2016; Das 2017). On the other hand, hemicellulose part is composed of xylan, pentosan, etc., and it is amorphous by nature. However, hemicellulose part of the jute fibre has a tendency to degrade in the presence of alkali and high temperature.

Generally, natural cellulosic fibres have the advantage of reinforcing material over synthetic fibres. Other than reinforcement material, natural fibres also have advantages like low cost, comparable specific strength, low density, lower pollutant emissions, acoustic properties, lower energy requirement, wide availability, renewable natural resource, etc (Ammayappan et al. 2016; Das 2017; Teli and Pandit 2017b).

In the internal structure of bast fibres from plant source, the microfibrils are well aligned with respect to the fibre axis. In woody plant tissue, the embedding matrix is reinforced by lignin. Multiple shortcomings such as incompatibility with hydrophobic resins, high moisture absorption, poor wettability, low-dimensional stability, ageing rapidly at moderately high temperature limit its usage in high-tech applications. To overcome these problems, there is a need for some physical, chemical and surface treatments as described in the literature (Das 2017; Das and Bhowmick 2015).

2 Composites

Composites are built materials which are made up of two or additional constituents with completely different properties (physical or chemical) that shows difference among the finished structure. The composite could even have properties that surpass the properties of the individual constituents. Composites of assorted types surround us in everyday life, natural and manmade. Examples of natural composites are the human bones and wood. Nature is good in its construction of materials appropriate for various applications. Humans have used the thought of composite materials for ages in numerous applications like building blocks from straw and clay, concrete bolstered with steel and polymers bolstered by numerous styles of fibres.

Fibre-reinforced plastics or composites have been commonly used for structural applications due to their excellent mechanical properties with respect to metals which they replace. While they were developed for aerospace structures in the 1950s, advanced composites mostly found their applications in automotive parts to building materials and speciality sporting goods to circuit boards. Mostly commercially available composites are manufactured using non-degradable polymeric resins such as epoxies and unsaturated polyesters in combination with high tensile strength fibres such as graphite, aramids and glass. These composites are primarily designed for applications where strength, stiffness and long-term durability are required. Most of the resins and fibres used in these composites are derived from petroleum-based sources.

With the increasing number of mass volume applications such as aerospace and civilian structures, green manufacturing and life-cycle assessments have made environmentally safe composites, a goal to work towards. Since composites are made using two dissimilar components that are bonded together and have a specific shape, they cannot be recycled or reused as of now more than 90% of the composites are discarded in landfills. In landfills, they may not degrade for a number of years or decades making that land unusable for any other application.

The ever-growing plastic and composite waste and pollution issue have raised the environment consciousness among the consumers, government alike and manufacturers. The assessment of the current rate of consumption of petroleum concluded that it is at an unsustainable rate of 100,000 times the rate at which it is created by the nature (Stevens 2002). To solve these pollution problems, government bodies established a law to encourage the utilization of recycled bio-based green products throughout the world (Nir et al. 1993). Nowadays, there is a rise in global environmental awareness, high rate of depletion of petroleum resources, new environmental regulations and concept of sustainability which triggered the need of new products and processes that are eco-friendly in nature (Netravali and Chabba 2003).

Sustainability, industrial ecology, 'cradle-to-cradle' design, green chemistry and eco-efficiency are the new principles that underlie the development of advanced green materials. Composite materials also will perforce follow the new paradigm. Most global manufacturers are marching towards green or recyclable products. Undoubtedly, with all the research that is currently going on at present, we can only expect that environment-friendly, sustainable and fully biodegradable green plastics and composites will be in every part of our life.

2.1 Sustainable Fibres Used in Composite

Polymer composites will contain fibres of various origins with numerous properties. The fibre choice is often driven by the necessities of the ultimate product. Mineral fibres are usually employed in applications like electrical insulators and boat hulls. Carbon fibres that most frequently are made from PAN fibres are applied in aerospace, sports merchandise and so forth. Glass, carbon and alternative standard reinforcement materials are out there as continuous fibres in roving or materials of assorted varieties or sliced in mats. The properties of those reinforcements are well outlined and documented consistently. Natural fibres, here outlined as fibres from a natural resource (i.e. a plant or a tree), are not used for load-bearing structures to any massive extent nowadays. The fibres are sometimes solely a few millimetres up to a couple of centimetres long. However, these are often spun to roving and weaved to materials which may be used for structural applications.

However, these fibres have their limitations and downsides and it is vital to bear them in mind so as to use the fibres for the proper applications (Netravali and Chabba 2003; Bhattacharya and Misra 2004; Roy et al. 2009; McDowall et al. 1984):

- Hydrophilic in nature.
- The compatibility between hydrophilic fibres and hydrophobic matrices is low.
- The fibres are not immune to high temperatures (>200 °C).
- The fibres are short.
- The standard and uniform consistency of properties is buffeted by factors that are exhausting to manage like climate throughout growth and harvest.

The main advantage for these fibres over all alternative forms of fibres is their origin. Plants and trees regenerate and that they are carbon dioxide neutral.

2.2 Necessity of Matrix and Its Role in Preparation of Fibre-Reinforced Polymer Composite

Generally, when composite is prepared from natural fibres, the matrix plays a great role of keeping the fibres in situ, distributes, and transfers stress among the fibres. It also provides a barrier against the hostile environment (chemicals and moisture) and gives surface protection to the fibres from mechanical degradation (e.g. by abrasion) (Bhattacharya and Misra 2004). The matrix plays a bit part in carrying the tensile load within the structure of composite. However, choice of a matrix casts a large impact on the compressive, lay-to-rest laminar shear along with in-plane shear properties of the material.

2.2.1 Resin Matrix

Composite materials are embedded with fibre in the polymer matrix. Generally, polymer is often classified into two categories, thermoplastic and thermosetting. Thermoplastic materials presently dominate, as matrices for biofibres, with the most frequently used thermoplastics for this purpose being polypropylene, polythene and polyvinyl chloride, whereas phenolic resin, epoxy and polyester resins are the most used thermosetting matrices.

2.2.2 Thermoset Benefits

• Thermal stability,
• Chemical and water resistance,
• Low creep and stress relaxation,
• Low viscosity which is excellent for fibre orientation.

2.2.3 Thermoplastic Benefits

• Room temperature storage of material,
• Economical,
• Reformable,
• Forming pressures and temperatures.

Thermoplastics soften when heated, and eventually, hardening starts to occur with cooling. This process of melting could be a continuous range instead of a point on the temperature scale with no considerable impact on materials properties. Thermoplastics embrace nylon 6 or 66, polypropylene, etc., which are often bolstered, with short, sliced fibres like a glass thermoset materials, are shaped in a in situ reaction, where the hardener and resin or the catalyst and resin are mixed, so it undergoes an irreversible reaction which creates tough product. Some thermosets, like phenolic resins, are made as side products (condensation reaction). Alternative thermosetting resins like epoxy and polyester are cured by mechanisms which do not manufacture volatile by-products, and they are much easier to process addition reactions. Thermosets will not melt once cured even if heat is applied, although their overall mechanical properties will change at that temperature considerably. This phenomenon is called as glass transition temperature (Tg), which varies with the actual resin system applied. The molecular structure and resin modulus properties change over Tg. Alternative properties like water and chemical resistance and colour stability conjointly scale back markedly at Tg. Though there are many alternative forms of resins which are used within composite industries, the bulk of structural elements mostly is created using three varieties, specifically polyester, vinyl ester and epoxy resins (Roy et al. 2009).

2.2.4 Polyester Resins

Generally, polyester resins are applied in resin systems, notably within marine industries, e.g. dinghies, work boats, yachts. This resin is of 'unsaturated' sort. The unsaturated polyester resin is thermosetting, means it has the capacity to be cured to solid form from liquid form when subjected to ideal conditions. The unsaturated resin differs from a saturated resin like Terylene™ that cannot be cured during this approach. Unsaturated polyester resins are referred to as polyesters. When reaction between organic acid and alcohol takes place, an ester and water are produced. Reaction of glycol with di-basic acids produces polyester and water. In composite industry, two types of polyester resin are used as a customary laminating system. Orthophthalic polyester resin is an economical resin which is employed commonly. Isophthalic polyester resin currently is a popular product in marine industry where its excellent water resistance properties are put to use (Saikia and Ali 1999).

Polyester matrix, as a class, has the subsequent advantages:

- Economical,
- Long history of performance,
- Majorly applied in transportation, construction sites, marine companies.

2.2.5 Vinyl Ester Resins

If vinyl ester resin is compared with polyester resins, then it was observed that they are almost similar in molecular structure. These resins conjointly feature less ester groups, and these ester groups are susceptible to water degradation by chemical reaction which suggests that vinyl esters exhibit higher resistance to water and lots of alternative chemicals than their polyester counterparts. Due to this property, vinyl esters are often used in applications like chemical storage tanks and pipelines. The fabric is thus usually used as a 'skin' coat or barrier for a polyester laminate that is to be immersed in water, like in a very boat hull. The molecular structure cured by vinyl organic compound makes it harder than polyester (Saikia and Ali 1999).

2.2.6 Epoxy Resins

Epoxies usually overshadow most of the alternative organic compound varieties in terms of resistance to environmental degradation and mechanical properties that end up in their virtually exclusive applications in craft elements. It is used as coating organic compound where their exaggerated resistance to water degradation and adhesive properties makes these resins ideal to be used in applications like boat manufacturing. Epoxy resins are usually used in the construction of superior boats or as a secondary application to interchange water-degraded polyester resins. Liquid resins and therefore the hardening agents form low viscosity simply processed systems. These resins are quickly and easily cured at temperature from 5 to 150 °C and are a favourable option in the selection of hardening agent. Epoxy is used in sealing products, paints and varnishes, coating products, etc.

The matrix prepared from epoxy, as a class, has the subsequent advantages:

- Wide selection of properties, since an outsized range of beginning materials, modifiers and hardening agents are out there.
- Throughout cure, there is absence of volatile matters.
- Low shrinkage throughout cure.
- Good resistance to solvents and chemicals.
- Excellent adhesion to a large form of fibres, fillers and alternative substrates.
- Major uses are strengthening systems, stay-in-place forms.

Some of the composites consisting of novel rural fibres are tested for their properties and potential to be employed in industrial material.

3 Surface Modification of Bast Fibres for Composite Materials

Jute, flax and ramie are the most used lignocellulosic bast fibres for composite applications. The three basic constituent components of bast fibres are cellulose, lignin and hemicelluloses. Cellulose is the major structural component of bast fibres (Kabir et al. 2012). Cellulose macromolecule is a semicrystalline polysaccharide made up of anhydro-D-glucose, which contains three alcohol hydroxyls. Hemicellulosic component is branched, fully amorphous and containing many acetyl and hydroxyl groups in their molecule. Lignins are highly complex, amorphous, polymers of phenyl propane units, mainly aromatic but have lesser water sorption than the cellulose and hemicelluloses. These hydroxyls groups are able to form hydrogen bonds with hydroxyl groups of water molecule present in the air. Therefore, all the bast fibres are hydrophilic in nature. The main limitation of cellulosic bast fibre is their hydrophilic nature, which reduces their compatibility with hydrophobic polymer matrices (Debnath et al. 2013). Bast fibre contains trace amount of waxy substance on their surface which affects the fibre–matrix bonding and surface wetting property. The presence of free water molecule in fibre reduces the adhesive characteristics with most binder resins. In the fibre–matrix interface, the water molecule performs as a separating agent. An optimum fibre–matrix interface bond is important to get high mechanical properties of composites (Drzal and Madhukar 1993; Bledzki et al. 1996).

3.1 Surface Modification Methods of Natural Fibres

Chemical and physical methods can be used to modify the fibre surface for better interface. To remove the attached organic compounds to the hydrophilic surface of plant cell walls of bast fibres, a number of surface modification techniques have been used.

3.2 Physical Methods

In this method, the chemical composition of the fibres does not change. The physical treatments change the fibre surface properties and structure which improve the mechanical bonding to polymers. Various types of physical surface modification methods are applied such as stretching, calendaring (John and Anandjiwala 2008; Thaman 1995), the production of blended yarns, thermotreatment (Ray et al. 1976; Zhang and Wadsworth 1999), corona treatment, UV treatment and plasma treatment. Corona treatment is one of the most widely used treatments for surface modification (Gassan and Gutowski 2000).

3.3 Chemical Methods

Several chemical methods are employed to make compatible the sustainable bast fibres with hydrophobic matrices. Alkaline treatment is one of the most important chemical treatments of bast fibres when used to reinforce thermoset resins.

3.3.1 Alkali Treatment

Alkali treatment though does not cause chemical modification, it reduces the amorphous content of the fibres by removing hemicellulose and wax substances and will increase their surface roughness, and thermal stability enhancing overall mechanical properties of the composites prepared.

Sodium hydroxide is a most typical chemical used for the processing of plant fibres. This results in the removal of lignin, hemicellulose, pectin, waxy substances and surface impurities. Consequently, it reveals the fibrils and offers a rough surface topography to the fibre. It conjointly transforms into cellulose II from cellulose I by a method called alkalization (Van de Weyenberg et al. 2006). Alkalization primarily causes swelling of cellulosic fibres, and therefore, the degree of swelling depends on the type of alkali used. It is found that Na^+ in NaOH has a most suitable diameter, ready to enlarge even the tiniest pores in between the cellulose lattice planes and allows them to penetrate to the most extent giving the highest degree of swelling (Oksman et al. 2003). After the removal of excess NaOH on washing, the new cellulose I lattice is created, with comparatively massive distances between the cellulose molecules, and these areas are stuffed with water molecules. This occurs wherever the –OH groups of the cellulose regenerate into –ONa groups, but continuous rinsing with water can take away the connected Na^+ ions and convert the cellulose to a new crystalline structure cellulose II, which is more stable than cellulose I. The reaction between fibre and NaOH is shown below.

$$Fibre - OH \ + \ NaOH \ \rightarrow Fibre - O - Na^+ + H_2O + Surface \ impurities$$

3.3.2 Graft Copolymerization

Synthetic change through graft copolymerization is an effective procedure of modifying the properties of fibres. The procedure includes the grafting of arranged monomers onto the surface of cellulosic strands. The reaction is normally started by free radicals on to cellulose. The cellulose is presented to high-vitality ionizing radiation. After treatment with selected ions, transition metal ions, oxidizing reagents, as initiators, creates free radicals on cellulose. The radical sites initiate grafting of vinyl monomers on the cellulose backbone.

Graft copolymerization is an ordinarily utilized approach for the modification of fibre surfaces (Kalia and Sabaa 2013), and it is a vital tool to change the physio-chemical properties of fibre or polymers.

Grafting attaches side chains covalently to most of the polymer backbone or polymer substrate to make a polymer with expanded chemical structure. Graft copolymers have numerous advantageous properties not quite the same as those that have no grafting.

Grafting is regularly performed in an exceptionally homogenized or in an extremely heterogeneous medium. Initiation reaction mechanism falls into three groups:

(a) The graft copolymerization of a vinyl monomer on polymer backbone by propagation reaction is initiated by chain transfer. In this manner, the development of graft chains starts due to the presence of active sites on the backbone of polymer. This phenomenon is named as 'grafting' method. In this methodology, the grafting is either performed with one monomer or monomer mixture blend (Zheng et al. 2010).
(b) The graft polymerization procedure of vinyl monomer in the presence of polymer with responsive reactive functional groups. These groups are regularly actuated by that of light, heat or other ways. This philosophy is frequently performed by two routes: 'grafting to' and 'grafting through' approaches (McDowall et al. 1984).
(c) The grafting of a vinyl monomer onto the polymer backbone by high-energy irradiation. Radiation grafting is regularly accomplished by coordinate/common procedure in which the polymer substrate and hence the monomer are irradiated to generate the radicals which are beginning the polymerization reaction or by pre-irradiation grafting amid which the monomer is in contact with the polymer backbone.

Contingent upon the chemical structure of vinyl monomer grafting onto cellulose backbone is the modulated properties. Enhanced properties like water assimilation, enhanced flexibility, hydrophilic or hydrophobic character, colour sorption abilities, thermosensitivity, pH scale sensitivity, antibacterial impact (McDowall et al. 1984; Doerffer 2013) in order to get a grafted cellulose with high water or wet porousness, hydrophilic monomers like acrylic acid, acrylamide, 2-acrylamidomethylpropane acid ($AASO_3H$) are grafted. Methyl methacrylate, styrene, acrylonitrile, butadiene, vinyl acetic acid and isobutyl divinyl ether are hydrophobic monomers (Margesin and Schinner 1999).

3.3.3 Grafting of Cellulose

Chemical grafting is amongst the most favoured ways for modifying the structure and properties of biopolymers. Graft copolymerization with various vinyl monomers onto cellulose backbone may be a method during which attempts are created to mix synthetic monomers with cellulose, to provide material with the best properties of each. In graft copolymerization, side chain grafts with purposeful groups are covalently bonded up to main chain of a chemical compound backbone to create branched polymer. By chemical modification of cellulose through graft copolymerization with synthetic monomers, many alternative properties, such as water permeability, elasticity, thermal resistance, processing capabilities and resistance to microbiological attack, are often improved (Hill et al. 1998). These grafted fibres were used in the preparation of sustainable composites. Modified fibres showed good tensile, flexular and impact properties of composites compared to that of unmodified.

3.3.4 Other Chemical Treatments

Silane is a chemical compound with chemical formula SiH_4 which can be used as a coupling agent for cellulosic fibres. These coupling agents are able to reduce the number of hydroxyl groups present in the cellulosic fibre (Jähn et al. 2002). Organic peroxides decompose easily to form free radicals of the form RO; RO then reacts with the hydroxyl group of the matrix and cellulose fibres. Benzoyl peroxide and dicumyl peroxide are generally applied in natural fibre surface modification purpose. Permanganate treatment is used for bast fibre surface modification purpose. Mostly, potassium permanganate solution is used for permanganate treatments of cellulosic fibres (Paul et al. 1997).

4 Application of Sustainable Raw Materials as Green Composite

Plant-based cellulosic fibres are mostly used for reinforcing non-degradable plastics to make greener composites. Their use has been fuelled by their being inexpensive and availability worldwide. These fibres are non-abrasive to processing equipment due to its hollow and cellular nature which perform well as acoustic and thermal insulators. Further, treatments such as mercerization and silanes have shown to enhance their mechanical properties and reduce their moisture sensitivity, respectively.

Many degradable plant-based fibres are used for reinforcing non-degradable thermoplastic polymers polyethylene, polyvinylchloride, nylons and polyesters to produce as 'greener' composites. Plant-based fibres have also been used with thermosetting resins such as epoxy and polyurethane (Kalita and Netravali 2017). Most of these composites, however, are made using sawdust, a waste product from saw mills, or wood fibre obtained by grinding wood products. These composites, are called 'wood–polymer composites' (WPCs) with wood fibre content ranging between 30 and 70%, are commonly used in ·non-structural applications (Wiener et al. 2003). The market for such composites has seen double-digit growth in the past decade and is expected to grow further in the near future. Wood fibres and flour are also used with formaldehyde-based resins to produce particle boards and medium density fibre boards (MDF) as wood. Longer plant-based fibres such as banana, jute, flax, kenaf, ramie, linen, henequen, hemp, sisal, pineapple have good mechanical properties. While they may not replace high strength fibres such as Kevlar®, graphite or glass, they can be used as low-cost alternative reinforcements for composites requiring moderate strength. One of the major advantages of these fibres, derived from plant stems or leaves, is that they are renewable annually as compared to wood, which takes 20–25 years to grow to full maturity. Some natural fibres, such as flax, pineapple, may have strength up to 1000 MPa depending on the processing and other conditions. Other fibres such as ramie can have modulus of over 120 GPa. These fibres with their density in the range of 1.4–1.5 g/cc can have specific mechanical properties comparable to E-glass fibres. The hollow structure of some plant-based fibres also provides better insulation against noise and heat which can be used in sound absorption and insulators. Because of their noise-absorbing characteristics, these composites should also work well for walls of cubicles commonly seen in offices and call centres. Besides natural fibres, regenerated cellulose fibres such as high tenacity viscose rayon may also be used as reinforcement in green composites. While the conventional viscose rayon production is a complex and polluting process, recently developed Tencel^{TM} fibres offer a non-polluting method (Chand and Hashmi 1993). Other significant advantage of these fibres is their availability in continuous form compared to plant-based fibres which can only be obtained in short lengths (Netravali and Pastore 2014).

Some plant-based oils are chemically modified in order to produce inexpensive resins for 'greener' composites of different class. Soybean oil-based resins were first used by Henry Ford in 1938, whereas epoxy-based castor oil is used in crosslinked resin systems for many years; many other oils can be epoxidized as well. Although derived from biodegradable resource, once crosslinked, these resins become non-degradable 'to petroleum-based resins'. These 'greener' composites combining non-degradable resins with degradable fibres suffer from the same problem of disposability. There are some efforts to burn these composites to recover energy.

4.1 Sustainable Green Composites from Natural Oil-Based Resins

In 2001–2003, these institutes proposed the existence and applications of a new type of resin derived from natural oils or triglycerides and subsequently claimed them as an alternative of petroleum-based resins. Natural oils are excellent sources of long-chain unsaturated fatty acids (usually 0–3 double bonds per triglycerides) joining at a glycerol juncture. In most cases, chain lengths of these fatty acids mostly vary from 14 to 22 carbons. The double bonds or the allylic carbons or ester groups, and the alpha carbon atoms which are adjacent to the ester groups, are highly amenable to chemical reactions and are demonstrated as the active sites in the triglycerides. These active sites can be used to introduce various polymerizable structures within the chain, whereas the employed synthetic techniques remained similar. Therefore, these synthetic pathways are the enabling factors to accomplish higher level of molecular weight, crosslink density as well as to incorporate novel chemical functionalities (e.g. aromatic or cyclic structures) for imparting stiffness in a polymer network. Either by utilizing or simply functionalizing the various fatty acids, one can synthesize various types of resins like polyester resin, epoxy resin, polyurethane and vinyl resin. This pathway simply brought about a revolution in green composite-related research. Following this, the scientific world observed a rapid expansion of this area through research. Frequent reports are published about different oil-derived different resins and their applications in various fields, e.g. coating, composite and plasticizer. As these bioresins possess low cross-linking density, molecular weight and are incapable of displaying ideal rigidity and strength required for structural applications, they ultimately showed few disadvantages and failed to be utilized alone (Grishchuk and Karger-Kocsis 2011).

Therefore, manufactures or researchers have tried to blend them with other synthetic materials. Despite this disadvantageous property, such resin-based composite structures have been claimed to be green composite. Modification procedures regarding the enhancement of such properties are also available but later discouraged due to the utilization of soybean oil which has high food value. A few groups also reported the green composite preparation procedures employing quite costly linseed oil or castor oils. In spite of several reports, application of jojoba oil and canola oil lagged behind in usage because of their low availability (Jeske et al. 2007).

So far, a lot of research work has been performed worldwide on the use of natural fibres for the preparation of various types of composites as a reinforcing material. Advantages of natural fibres, which are used as sustainable raw materials, arise from their properties such as low density, ease of preparation, easy availability, biodegradability, renewability, low cost, lower energy requirements for processing, and relative non-abrasiveness over traditional reinforcing synthetic fibres. Moreover, natural fibres are environmentally friendly and neutral with respect to CO_2.

4.2 Sustainable Green Composites from Soy Protein-Based Resin

Soy proteins are very complex macromolecules usually possessing about 20 amino acids having many reactive sites which are available for the interaction in the presence of plasticizer. Hence, extrusion method is used to convert soy protein plastic with additives like a plasticizer or cross-linking agent. Two main varieties of soybean protein, available in the market, are soy protein isolate containing 90% protein and soybean protein concentrate (SPC) containing 50–70% protein. The soy proteins crosslink through covalent sulphur crosslinks under oxidative conditions at cysteine residues. Dehydroalanine (DHA), formed from alanine by the loss of side chain beyond the β-carbon atom, also reacts with lysine and cysteine to form lysinoalanine and lanthionine crosslinks, respectively (Lodha and Netravali 2002). Besides, asparagines and lysine can also react together to form an amide type of crosslink. All these reactions occur during the curing process of the SPI polymer forming a resin of moderate strength. The SPI polymer is highly hygroscopic/ moisture sensitive, due to the presence of polar amine, amide, carboxyl and hydroxyl groups (Nzioki 2010). The overall mechanical properties of soybean protein plastic are controlled by varying the moulding temperature, pressure and initial moisture content (Liang et al. 1999). Soy protein plastic is often blended with biodegradable polymer to form soy protein-based biodegradable plastic (Grishchuk and Karger-Kocsis 2011; Mohanty et al. 2005). Fibre-reinforced composites will find increasing use of these materials with their concurrent application in different sectors such as automotive industries and packaging products. The natural fibres used in such composites are not only eco-friendly and sustainable but also low cost with good overall mechanical properties.

4.3   Sustainable Green Composites from Polylactic Acid-Based Resin

Polylactic acid is a biodegradable polymer which is manufactured from biotechnological processes using renewable resources, as the raw material. Corn is mainly used as the starting raw material. Corn produces highly purified lactic acid. Apart from corn, other raw materials such as woody biomass are chosen in order to reduce the overall processing costs. Pure lactic acid which is derived from such raw materials is very cost effective at present. Lactic acid is used to synthesize PLA. There are usually two main methods, i.e. direct polycondensation and ring-opening polymerization of lactate monomer in the presence of a catalyst. Lewis acid is used as the catalyst. The largest producer of PLA is Cargill Dow LLC, which has also patented a low-cost continuous process for the production of polylactic acid polymer (Avérous 2004). PLA has very unique and interesting properties such as excellent strength, transparency of film, biocompatibility and biodegradability. Properties such as morphology, crystallinity, thermal stability and other properties are hugely dependent on the molecular weight. The amorphous nature of PLA can be overcome by modifying PLA with the addition of plasticizers and forming a copolymer.

These properties of PLA are useful for the preparation of biocompatible devices; further advantage is that during secondary operations there is no need to replace the implant when bioresorbable PLA-based materials are used. PLA is known for its good strength and good stiffness. It can be used as plastic bags, sanitary products, diapers, disposable cups, plates and other medical applications.

69177

4.4 Sustainable Green Composites from Starch-Based Materials

Starch is a macromolecule made up of α-D-glucose units that is available in two forms. One is amylose molecule that is linear, and the other one is amylopectin with highly branched structure. The size and shape of starch granules depend upon the source. Starch can be used as a replacement to the synthetic polymers where composites do not require long-term durability. Mostly, the starch is derived from wheat, tapioca, potato, maize, etc. Starch is mostly semicrystalline in nature, and the degree of crystallinity depends on the amylopectin content. Starch is subjected to destructurization process which leads to thermoplastic material. Acetylation of starch can improve moisture resistance. Blends of starch with other polymers allow starch-based polymers with greater flexibility to be produced. The hydrophilicity and poor mechanical properties of starch have contributed to few applications in which starch is used singly and not as part of a blend (Avérous 2004).

5 Some Novel Sustainable Lignocellulosic Fibres Used as Reinforcing Materials

5.1 Fibres from Agave angustifolia Plant

One of the most fibres used here during this work is Agave angustifolia (AA). Agave angustifolia v. marginata is usually known as 'Banded Carribean Agave'. Agave angustifolia is a species that belongs to Agavaceae family. This plant is attractive and rugged, with proportionate arrangement, slim and rigid bayonet leaves generously margined with off-white colour. These rosettes are often 1 m in circumference, it has leaves around 50–80 cm long, and it has a long terminal spine at the tip of the leaf. It has spread as a decorative plant worldwide. Native from Costa Rica to North American nation and North America, the peak of the plant will reach up to 3–4 feet. Every rosette produces offsets round the lower plant base. Each plant produces 20–30 offsets spreading to 15 feet from the parent plant. It grows very quickly as it is somewhat tropical plant; however, it will not tolerate abundant frosting. It grows best in full sunshine; however, it will adapt to some shade. Caribbean Agave plant may be a hardy survivor, tolerating heat, drought and salty sea coast conditions (Thaman 1995). The raw fibre is composed of cellulose 66%, hemicelluloses 23%, lignin 6.5%, ash 2% and extractives 2.5% (Fig. 2).

Fig. 2 *Agave angustifolia* plant, fibre and SEM image

5.2   *Fibres from* Abelmoschus manihot *Plant*

Abelmoschus manihot is a flowering plant in the bush family Malvaceae. It is domestically named as 'Ranbhendi' in geographical area, India. It is a species of hibiscus, but is currently classified in the genus Abelmoschus. This can be a perennial vascular plant, growing up to 2 m tall. The leaves are broad and 10–40 cm long. The flowers are 5–8 cm in diameter, with five yellowish white petals, and have a reddish purple spot at the end of every floral leaf. The fruit is in capsule shaped, 5–20 cm long, containing several seeds. The ringlet is four-lobed, with unequal lobes. Its medicative properties are known (Eggli and Hartmann 2002). The raw fibre is composed of cellulose 45%, hemicelluloses 30%, lignin 14%, ash 2% and extractives 3% (Fig. 3).

5.3   *Fibres from* Sansevieria roxburghiana *Plant*

Sansevieria roxburghiana (S.R.) Schult and Schult. F. (Agavaceae) is a herbaceous perennial plant. It occurs in eastern coastal region of India, Indonesia, Sri Lanka and tropical Africa (Prakash et al. 2008). In India, the plant leaves have been historically used in medicine as a cardiotonic, medicative tonic in organ enlargement and rheumatism (Teli and Jadhav 2016a; 2017). The raw fibre of this plant consists of 54% cellulose, 30% hemicelluloses, 12% lignin, 2% ash and 2% extractives (Fig. 4).

Fig. 3 *Abelmoschus manihot* plant, fibre and SEM image

Fig. 4 *Sansevieria roxburghiana* plant, fibre and SEM image

5.4   *Fibres from* **Pandanus odorifer** *Plant*

Pandanus odorifer (P.O.) is an aromatic monocot species of plant in the Pandanaceae family, native to Polynesia, Australia, South Asia (Andaman Islands) and the Philippines and also occurs in the wild in southern India and Burma. Extracted *P. odorifer* raw fibre is composed of cellulose 44%, hemicelluloses 30%, lignin 21%, ash 2.5% and extractives 2.5% (Teli and Jadhav 2016b) (Fig. 5).

Fig. 5 *Pandanus odorifer* plant, fibre and SEM image

Scanning electron microscopy (SEM) showed the morphology of *A. angustifolia*, *A. manihot*, *S. roxburghiana* and *P. odorifer* fibres. Like other bast fibres, all the three fibres obtained have a thick layer of deposits on the fibre surface composed of lignin, hemicellulose, wax and pectin which protects cellulose fibres in the core. The XRD analysis showed that the fibres have crystallinity of 55% for AA fibre, 63% for AM fibre, 72% for SR fibre and for PO fibre it was 46.5% which is almost similar to that of other commercially available natural fibres. TGA analysis of the fibres had a very good thermal stability up to 270 °C. The tensile strength of three fibres was also good. These properties show that the fibres have a great potential to be used in composites. Although these plants grow in wild, if these fibres get commercialized, then the farmers could also get a new source of income by cultivating these plants on a large scale, as they require minimal maintenance and are easy to grow. The tensile properties, flexural and impact strength of the *A. angustifolia*, *A. manihot* and *S. roxburghiana* fibres showed comparable values with reinforcing fibres used today (Teli and Jadhav 2016a, b; 2017; Van der Geer et al. 2000). The chemical resistance of alkali-treated fibre composites from these plant fibres are significantly improved when subjected to corrosive chemicals, as compared to raw fibre composite; it can be applied with constructing acid/water storage tanks, tabletops, automotive frames and in building construction. The fibre/epoxy composites have also demonstrated good mechanical and chemical resistance properties.

6   Challenges in Sustainable Composites

The main challenge primarily lying in manufacturing green composites is that it should withstand high temperatures by eliminating the smell developed by natural fibres like kenaf, hemp, jute. The price of the natural fibre-based composites is very high which is not acceptable by the humungous automotive industries. Incomplete information regarding the performance of green and sustainable composites which is due to excessive variety of constituents is mostly challenging to overcome the barrier in this sector. The most important challenge is to select proper processing conditions and parameters, accurate selection of composite constituents, and overall material characteristics should be assessed in a process of achieving good-quality composites. Poor natural fibre properties are a major drawback which limits their application in green composites. Other challenges which imply in many green composite are that the material origin has to be minor influence on the whole life-cycle environmental impact. Product energy consumption, durability and service efficiency are often the main parameters for sustainability (AL-Oqla and Omari 2017). End-of-life treatment of disposed composite materials must be considered in the product design phase. Plastic waste in the environment has been recognized as an environmental problem. Leakage of residues and additives from plastic consumer products has been detected which has raised consumer concern. Bio-based plastic raw materials have been introduced and have entered the market. A biomass-based resource does not necessarily mean carbon dioxide neutrality. In our society, plastics are more or less unavoidable materials in many areas, such as: health care, food distribution, building material, cars and transportation, packaging home electronics. Environmentally sound and technically efficient material recycling must be introduced especially in high population countries such as India, China, many African countries.

7   Conclusions

Nature has granted an abundance of resources that are sustainable, replenishable, healthy and most critically eco-friendly. The discovery of these resources and processes to use them optimally, as in green composites, is the current need. Bast fibres extracted from the stem of the wild plants exhibit high tensile strength and low extensibility mostly used for making ropes, yarn, paper and composite materials. It was found that the chemical composition of a fibre affects its appearance, structure, properties and processability which play a crucial role for composite application. Much research in this direction either by modification of fibres or modifying the resins by physical and chemical methods to utilize the bast fibres as green composite materials is ongoing. Newer fibre sources have also been researched for their application in the preparation of hybrid green composites and for use in advanced technical textiles.

References

AL-Oqla FM, Omari MA (2017). Sustainable biocomposites: challenges, potential and barriers for development. In: Green biocomposites, pp 13–29

Ammayappan L, Das S, Guruprasad R, Ray DP, Ganguly PK (2016) Effect of lac treatment on mechanical properties of jute fabric/polyester resin based biocomposite

Avérous L (2004) Biodegradable multiphase systems based on plasticized starch: a review. J Macromol Sci Part C Polym Rev 44:231–274

Baley C (2002) Analysis of the flax fibres tensile behaviour and analysis of the tensile stiffness increase. Compos Part A Appl Sci Manuf 33:939–948

Basak S, Samanta KK, Chattopadhyay SK, Pandit P, Maiti S (2016) Green fire retardant finishing and combined dyeing of proteinous wool fabric. Color Technol 132:135–143

Bhattacharya A, Misra BN (2004) Grafting: a versatile means to modify polymers: techniques, factors and applications. Prog Polym Sci 29:767–814

Bledzki AK, Gassan J (1999) Composites reinforced with cellulose based fibres. Prog Polym Sci 24:221–274

Bledzki AK, Reihmane S, Gassan J (1996) Properties and modification methods for vegetable fibers for natural fiber composites. J Appl Polym Sci 59:1329–1336

Chand N, Hashmi SAR (1993) Mechanical properties of sisal fibre at elevated temperatures. J Mater Sci 28:6724–6728

Das S (2017) Mechanical properties of waste paper/jute fabric reinforced polyester resin matrix hybrid composites. Carbohydr Polym 172:60–67

Das S, Bhowmick M (2015) Mechanical properties of unidirectional jute-polyester composite. J Text Sci Eng 5:1

Das M, Chakraborty D (2008) Evaluation of improvement of physical and mechanical properties of bamboo fibers due to alkali treatment. J Appl Polym Sci 107:522–527

Das S, Bhowmick M, Chattopadhyay SK, Basak S (2015) Application of biomimicry in textiles. Curr Sci 109:893–901

Debnath S, Nguong CW, Lee SNB (2013) A review on natural fibre reinforced polymer composites. World Acad Sci Eng Technol 1123–1130

Doerffer JW (2013) Oil spill response in the marine environment. Elsevier

Drzal LT, Madhukar M (1993) Fibre-matrix adhesion and its relationship to composite mechanical properties. J Mater Sci 28:569–610

Eggli U, Hartmann HEK (2002) Illustrated handbook of succulent plants: dicotyledons. Springer Science & Business Media

Gassan J, Gutowski VS (2000) Effects of corona discharge and UV treatment on the properties of jute-fibre epoxy composites. Compos Sci Technol 60:2857–2863

Grishchuk S, Karger-Kocsis J (2011) Hybrid thermosets from vinyl ester resin and acrylated epoxidized soybean oil (AESO). Express Polym, Lett, p 5

Hill CAS, Khalil HPSA, Hale MD (1998) A study of the potential of acetylation to improve the properties of plant fibres. Ind Crops Prod 8:53–63

Huda S, Reddy N, Karst D, Xu W, Yang W, Yang Y (2007) Nontraditional biofibers for a new textile industry. J Biobased Mater Bioenergy 1:177–190

Jähn A, Schröder MW, Füting M, Schenzel K, Diepenbrock W (2002) Characterization of alkali treated flax fibres by means of FT Raman spectroscopy and environmental scanning electron microscopy. Spectrochim Acta Part A Mol Biomol Spectrosc 58:2271–2279

Jeske RC, DiCiccio AM, Coates GW (2007) Alternating copolymerization of epoxides and cyclic anhydrides: an improved route to aliphatic polyesters. J Am Chem Soc 129:11330–11331

John MJ, Anandjiwala RD (2008) Recent developments in chemical modification and characterization of natural fiber-reinforced composites. Polym Compos 29:187–207

Kabir MM, Wang H, Lau KT, Cardona F (2012) Chemical treatments on plant-based natural fibre reinforced polymer composites: an overview. Compos Part B Eng 43:2883–2892

Kalia S, Sabaa MW (2013) Polysaccharide based graft copolymers. Springer

Kalita D, Netravali AN (2017) Thermoset resin based fiber reinforced biocomposites. Text Finish Recent Dev Futur Trends 423–484

Kim JH, Lee SB, Kim SJ, Lee YM (2002) Rapid temperature/pH response of porous alginate-g-poly (N-isopropylacrylamide) hydrogels. Polymer (Guildf) 43:7549–7558

Kiruthika AV (2017) A review on physico-mechanical properties of bast fibre reinforced polymer composites. J Build Eng 9:91–99

Liang F, Wang Y, Sun XS (1999) Curing process and mechanical properties of protein-based polymers. J Polym Eng 19:383–394

Lodha P, Netravali AN (2002) Characterization of interfacial and mechanical properties of "green" composites with soy protein isolate and ramie fiber. J Mater Sci 37:3657–3665

Margesin R, Schinner F (1999) Biological decontamination of oil spills in cold environments. J Chem Technol Biotechnol 74:381–389

McDowall DJ, Gupta BS, Stannett VT (1984) Grafting of vinyl monomers to cellulose by ceric ion initiation. Prog Polym Sci 10:1–50

Mohanty AK, Khan MA, Sahoo S, Hinrichsen G (2000) Effect of chemical modification on the performance of biodegradable jute yarn-Biopol® composites. J Mater Sci 35:2589–2595

Mohanty AK, Tummala P, Liu W, Misra M, Mulukutla PV, Drzal LT (2005) Injection molded biocomposites from soy protein based bioplastic and short industrial hemp fiber. J Polym Environ 13:279–285

Mwaikambo LY, Ansell MP (2002) Chemical modification of hemp, sisal, jute, and kapok fibers by alkalization. J Appl Polym Sci 84:2222–2234

Netravali AN, Chabba S (2003) Composites get greener. Mater Today 6:22–29

Netravali AN, Pastore CM (2014) Sustainable composites: fibers, resins and applications. DEStech Publications, Inc.

Nir MM, Miltz J, Ram A (1993) Update on plastics and the environment: progress and trends. Plast Eng 49:75–93

Nzioki BM (2010) Biodegradable polymer blends and composites from proteins produced by animal co-product industry

Oksman K, Skrifvars M, Selin J-F (2003) Natural fibres as reinforcement in polylactic acid (PLA) composites. Compos Sci Technol 63:1317–1324

Paul A, Joseph K, Thomas S (1997) Effect of surface treatments on the electrical properties of low-density polyethylene composites reinforced with short sisal fibers. Compos Sci Technol 57:67–79

Prakash JW, Raja RD, Anderson NA, Williams C, Regini GS, Bensar K, Rajeev R, Kiruba S, Jeeva S, Das SSM (2008) Ethnomedicinal plants used by Kani tribes of Agasthiyarmalai biosphere reserve, Southern Western Ghats

Ray PK, Chakravarty AC, Bandyopadhaya SB (1976) Fine structure and mechanical properties of jute differently dried after retting. J Appl Polym Sci 20:1765–1767

Reddy N, Yang Y (2005a) Properties and potential applications of natural cellulose fibres from cronhusks. Green Chem 7:190–195

Reddy N, Yang Y (2005b) Structure and properties of high quality natural cellulose fibers from cornstalks. Polymer 46:5494–5500

Robson D (1993) Survey of natural materials for use in structural composites as reinforcement and matrices. Biocomposites Centre, University of Wales

Roy D, Semsarilar M, Guthrie JT, Perrier S (2009) Cellulose modification by polymer grafting: a review. Chem Soc Rev 38:2046–2064

Saikia CN, Ali F (1999) Graft copolymerization of methylmethacrylate onto high α-cellulose pulp extracted from *Hibiscus sabdariffa* and *Gmelina arborea*. Bioresour Technol 68:165–171

Stevens ES (2002) Green plastics: an introduction to the new science of biodegradable plastics. Princeton University Press, Princeton

Teli MD, Jadhav AC (2016a) Effect of alkali treatment on the properties of *Agave angustifolia* v. marginata fibre. Int Res J Eng Technol 3:2754–2761

Teli MD, Jadhav AC (2016b) Extraction and characterization of novel lignocellulosic fibre. J Bionanosci 10:418–423

Teli M, Jadhav A (2017) Determination of chemical composition and study on physical properties of *Sansevieria roxburghiana* lignocellulosic fibre. Eur J Adv Eng Technol 4:183–188

Teli MD, Pandit P (2017a) Novel method of ecofriendly single bath dyeing and functional finishing of wool protein with coconut shell extract biomolecules. ACS Sustain Chem Eng. https://doi.org/10.1021/acssuschemeng.7b02078

Teli MD, Pandit P (2017b) Development of thermally stable and hygienic colored cotton fabric made by treatment with natural coconut shell extract. J Ind Text 1528083717725113

Teli MD, Pandit P, Basak S (2018) Coconut shell extract imparting multifunction properties to ligno-cellulosic material. J Ind Text 47(6):1261–1290

Thaman RR (1995) Urban food gardening in the Pacific Islands: a basis for food security in rapidly urbanising small-island states. Habitat Int 19:209–224

Van de Weyenberg I, Truong TC, Vangrimde B, Verpoest I (2006) Improving the properties of UD flax fibre reinforced composites by applying an alkaline fibre treatment. Compos Part A Appl Sci Manuf 37:1368–1376

Van der Geer J, Hanraads JAJ, Lupton RA (2000) Clean energy project analysis: Retscreen® engineering & cases textbook, small hydro project analysis chapter. J Sci Commun 163:51–59

Van Voorn B, Smit HHG, Sinke RJ, De Klerk B (2001) Natural fibre reinforced sheet moulding compound. Compos Part A Appl Sci Manuf 32:1271–1279

Wiener J, Kovačič V, Dejlová P (2003) Differences between flax and hemp. AUTEX Res J 3:58–63

Zhang D, Wadsworth LC (1999) Corona treatment of polyolefin films—a review. Adv Polym Technol 18:171–180

Zheng L, Dang Z, Zhu C, Yi X, Zhang H, Liu C (2010) Removal of cadmium (II) from aqueous solution by corn stalk graft copolymers. Bioresour Technol 101:5820–5826

Production of Sustainable Green Concrete Composites Comprising Industrial Waste Carpet Fibres

Hossein Mohammadhosseini and Mahmood Md. Tahir

Abstract Green production and sustainable development are the pressing concerns of the twenty-first century. By the growing amount of wastes generated from many practices, there has been a growing consideration in the utilisation of wastes in the production of sustainable composite materials such as concrete composites to attain possible benefits. In the construction industries, the idea of sustainability inspires the usage of waste materials to replace raw resources, such as cement, aggregates and fibrous materials. This leads to sustainable, green and eco-friendly construction by decreasing the cost and natural resources related to disposing of the waste materials. This chapter highlights the outcomes of an experimental examination on the performance of green concrete composites comprising industrial waste polypropylene carpet fibres and palm oil fuel ash (POFA). Six different concrete mixes containing 20-mm-long carpet fibre at dosages of 0–1.25% were made with ordinary Portland cement (OPC). Additional six concrete mixes containing 20% POFA with the same fibre dosages were made. Concrete composite specimens then were tested for fresh and hardened state properties. It has been found that by adding carpet fibres the workability of composite mixes was reduced. Besides, the compressive strength of concrete composites did not improve by the addition of carpet fibres. However, the positive interaction amongst carpet fibre and POFA leads to lesser drying shrinkage, higher flexural and tensile strengths, thereby enhancing the ductility of concrete composites with higher energy absorption and impact resistance and, therefore, developed crack distribution. The water absorption and depth of chloride penetration of the concrete composites were significantly reduced with the inclusion of waste carpet fibre. The results further reveal that the replacement of POFA significantly influenced on the durability characteristics of concrete composites at extended curing times. The outcomes of this study proposed that the employment of industrial

H. Mohammadhosseini (✉) · M. Md. Tahir
Institute for Smart Infrastructure and Innovative Construction (ISIIC),
School of Civil Engineering, Faculty of Engineering,
Universiti Teknologi Malaysia (UTM), 81310 Skudai, Johor, Malaysia
e-mail: mhossein@utm.my

M. Md. Tahir
e-mail: hofa2018@yahoo.com

© Springer Nature Singapore Pte Ltd. 2019
S. S. Muthu (ed.), *Green Composites*, Textile Science and Clothing Technology,
https://doi.org/10.1007/978-981-13-1969-3_2

waste carpet fibre together with POFA in the production of sustainable and green concrete composites is viable technically and environmentally.

Keywords Green concrete composites · Sustainability · Waste carpet fibres Palm oil fuel ash · Physical and mechanical properties

1 Introduction

1.1 General Appraisal

Through industrialisation and technological developments in various fields, huge amount and different sorts of solid waste materials have been generated by the industrial, mining, agricultural and domestic actions. Therefore, waste management has become one of the main environmental anxieties all around the world. With the growing attentiveness about the environment, lack of landfill area and because of its high cost, utilisation of by-products and waste materials has become an attractive substitute for discarding. Recycling of the non-biodegradable wastes is very difficult. Utilisation of natural sources, massive quantity production of industrial waste and environmental contamination need gaining new and applicable solutions for sustainable development. Over the decades, there has been a rising affirmation of the use of by-products and waste materials in the construction industry. Utilisation of the wastes not only aids in getting them applied in concrete composites, and other similar applications, it benefits in decreasing the cost of the concrete producing, but also has many indirect advantages such as a decrease in landfill area, saving in energy and defending the environment from harmful impacts. Moreover, consumption of these waste materials may enhance the physico-mechanical properties, durability performance and the microstructure of concrete composites, which are challenging to attain by the usage of only raw materials (Putman and Amirkhanian 2004; Batayneh et al. 2007; Meddah and Bencheikh 2009; Kanadasan and Abdul Razak 2015; Gu and Ozbakkaloglu 2016).

A primary challenge facing the construction industries is to execute projects in compatibility with the environment by adopting the concept of sustainable growth. This includes the usage of high-performance and eco-friendly materials manufactured at a reasonable quality and cost. Current researches on many waste materials such as supplementary cementing materials (SCM), plastics and textiles, aggregates and a host of others have shown that the addition of such waste materials in concrete has the potential to enhance the physical, mechanical and durability of concrete as well as a reduction in the cost of construction (Chandra 1997; Siddique et al. 2008; Thomas and Gupta 2013). The challenges are more a consequence of the facts that Portland cement is not particularly eco-friendly and lack of landfill space for waste materials. One could then decrease these challenges to the succeeding simple formulation: use as much concrete, but with as low OPC as possible, and waste materials as much as possible, this means to substitute as much raw material as possible by waste and SCMs, particularly those that are by-products of industrial processes, and to use wastes instead of raw materials.

Synthetic fibres are industrialised mainly to supply the high demand for carpet and textile products where polypropylene and nylon are the most used synthetic fibres. Generally, carpets are categorised as textile and waste carpets are mainly caused both from post-consumer (old carpets) or pre-consumer (industries). The approximate amount of the industrial waste carpet fibres generated in Malaysia is estimated as 30–50 tons annually, reported by Malaysian carpet industries. The aids of using such waste fibres include the lesser cost to prepare than fibres available in the market, lighter in weight, superior resistance to chemicals and hydrophobic in nature (Wang et al. 1994, 2000; Schmidt and Cieslak 2008; Ghosni et al. 2013; Awal and Mohammadhosseini 2016). The pozzolanic ashes are employed in construction industries for their advantages. Palm oil fuel ash is one of the latest inclusions in the ash family which is attained by incineration of palm oil husk and kernel shell as fuel in palm oil mill (Tay 1990; Awal and Hussin 1997; Tangchirapat et al. 2007). In 2007, about 3 million tons of POFA had been produced in Malaysia only, and this manufacturing rate is likely to increase owing to growing the plantation of palm oil trees (Ismail et al. 2011; Al-Mulali et al. 2015; Ranjbar et al. 2016; Mohammadhosseini et al. 2017).

1.2 Background

Concrete is the essential construction materials, and its utilisation is rising all around the globe. Besides the regular applications, higher energy absorption capacity and ductility are mostly required in various fields like industrial buildings, highway pavements, bridge decks. Nevertheless, conventional concrete possesses very slight tensile strength, limited ductility, low resistance to cracking and little energy absorption. Internal microcracks inherently exist in the concrete specimens, and its low tensile strength is owing to the propagation of such microcracks, ultimately leading to brittle fracture of the concrete. Therefore, enhancing the toughness of concrete and decreasing the size and possibility of weaknesses would lead to better concrete performance.

Previously, efforts have been made to impart enhancement in ductility and tensile strength of concrete with the addition of a small fraction (0.5–2%) of short fibres to the concrete mixture throughout mixing process (Zollo 1997; Brandt 2008; Yahaghi et al. 2016). In such situations, fibre-reinforced concrete (FRC) has been shown to perform its purposes adequately. FRC can be described as a composite material containing mixtures of cement and binders, coarse and fine aggregates and short fibres that are dispersed in the concrete matrix. There are various types of fibres, no matter polymeric or metallic, utilised in FRC for their benefits. Amongst others, the most common types of fibre used in concrete composites are glass fibres, steel fibres, polypropylene (PP), natural fibres and fibres produced from wastes.

Fibres in general and polypropylene fibres, in particular, have gained popularity recently for use to improve the properties of concrete (Brandt 2008; Mohammadhosseini and Awal 2013).

In brittle materials like plain concrete without any fibre, microcracks develop even before applying a load, mainly owing to drying shrinkage or any other cause of volume variation. While loading, the cracks propagate and open up, owing to the effect of stress concentration and formation of additional cracks in places of minor defects. The development of such microcracks along the concrete members is the main reason of inelastic deformation in concrete (Hsie et al. 2008; Mohammadhosseini and Yatim 2017). It has been recognised that the addition of short PP fibre in a concrete mixture is potential in bridging the cracks, load transfer, and improving microcracks dispersal system (Aldahdooh et al. 2014). Moreover, the fibres would act as crack arrester and would significantly enhance the properties of concrete not only under compression, tensile and flexure (Yap et al. 2013), but also under impact blows (Nili and Afroughsabet 2010) and plastic shrinkage cracking (Zhang et al. 2011).

One of the fundamental solutions towards attaining enhanced concrete properties in terms of strength, durability and microstructures is the combined use of polypropylene fibre and pozzolanic materials in concrete. Polypropylene fibre is present in the mixture to reduce the brittleness of the matrix, thus reducing the susceptibility to cracking of a concrete (Karahan and Atis 2011). As most of the problems related to the durability properties such as permeability, chloride penetration, carbonation and acid and sulphate attacks start from concrete cracking, a considerable way that reduces the brittleness of concrete is necessary and foremost efficient. Fibre-reinforced cementitious composites address the brittleness of concrete. This ductile material exhibits a good ductility under mechanical loads and durability under severe environmental exposure (Yap et al. 2014; Mo et al. 2015; Mohammadhosseini et al. 2016).

2 Waste Carpet Fibres

World synthetic fibre manufacture has been gradually growing in the last decades, currently beyond 64 million tons annually. Generally, the uses of fibres belong to the following three extensive classes: clothing, home furnishing and industries. The useful life of the fibre products may vary from short range (e.g. disposables), medium term (e.g. apparel, carpet, automotive interior), to long standing (e.g. textiles for construction) (Wang 2010). Figure 1 displays the carpet waste management hierarchy according to Carpet Recycling UK (Bird 2013; Sotayo et al. 2015). It can be seen that the most favoured option is the prevention. It can be attained over the usage of less raw resources in the design and the production process, which, in turn, decreases the waste produced in addition to the greenhouse gas emissions related to the production of carpets (Department for Environment Food and Rural Affairs 2014). As shown in Fig. 1, prevention proposes the best

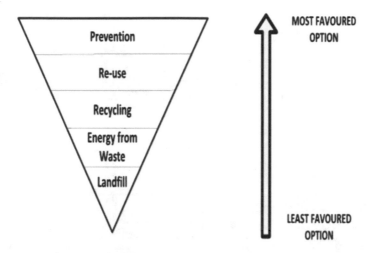

Fig. 1 Carpet waste management hierarchy

consequence for the environment, followed by reuse, recycling, energy from waste (by burning) and finally landfill disposal (Bird 2013; Sotayo et al. 2015).

In the UK, textiles account for about 2–5% of all wastes which are send to disposal. Based on the information published by the Carpet Recycling UK, approximately 400,000 tons of various sorts of waste carpet are led to landfills yearly (Sotayo et al. 2015; Mohammadhosseini et al. 2018). Similarly, roughly 1.9 million tons of waste textiles were made in the USA by 2007, which is counted as 4.7% of the total public solid wastes generated (Wang 2010). Industrial waste carpets are mostly from face and back yarns, where the face yarns are mainly nylon (50–70%) and polypropylene fibres (15–25%) (Schmidt and Cieslak 2008; Ucar and Wang 2011) and the back yarns are usually in the form of woven sheets as shown in Fig. 2.

2.1 Types and Sources of Waste Carpet

Waste fibres generated from post-consumer raises to products thrown away after their service lives. As nearly all the waste fibres are transformed into different forms and use as a substitution for worn out or outdated products, the quantity of post-consumer wastes is excessive and is equivalent to the rate of fibre intake. Old carpets or rugs, which are usually used as floor coverings, are composite materials that are difficult and costly to separate and reprocess at the end of their useful lives. It is because they are multilayer mixtures of different polymers and inorganic fillers. To support strong, profitable actions to reprocess post-consumer wastes, some assemblage and organisation complexes are necessary. There are recognised methods to reutilise old carpets and textiles. Moreover, to straight reuse of carpets,

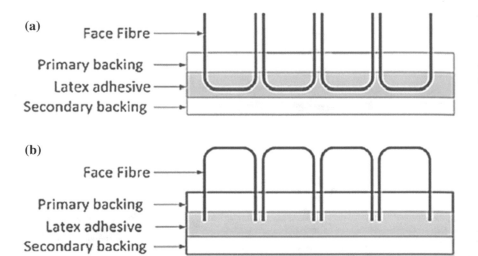

Fig. 2 Typical arrangement of industrial carpets: **a** cut pile and **b** level loop

waste composed could be transformed into wipes or be shredded for filling or
non-woven requests (Wang 2010). However, this kind of waste fibres contains a
significant amount of impurities that could negatively affect the properties of
concrete mixtures.

Another source of waste carpet fibres is pre-consumer, or those are generated
directly from the industry during the manufacturing process. The type of carpets is
categorised based on the type of fibres used in the face yarn. A nylon 6 type of
carpets, for example, comprises not only nylon 6 fibres in face yarn but also
contains PP fibres and adhesives such as latex and fillers at the back yarns (Wang
2010). Nylon fibre usually executes the best amongst all polymeric fibres as face
yarn of carpet. It is more expensive than polypropylene fibre. Therefore, due to its
higher price, wasting of this fibre is minimised during the manufacturing. Amongst
others, polypropylene fibres are the most wasted fibres in the manufacturing process
of industrial carpets. It is due to the lower cost of the PP fibres and also broad
applications of carpets made with this type of fibre all around the world.
Consequently, the primary concern of this research was on the utilisation of waste
polypropylene fibres from industrial carpets. A typical source of pre- and
post-consumer carpet waste fibres is illustrated in Fig. 3.

2.2 Carpet Fibre Recycling Technology

To help recycling industries that are accomplished of processing a substantial
quantity of the wastes disposed of, there should be a collection system to deliver an
adequate and reliable supply of pre- and post-consumer fibres at a sensible price and

Fig. 3 Typical pre- and post-consumer waste carpet fibres

a variety of active processes to produce a diversity of profitable product. Variety of waste fibres is an important issue for the achievement of the industries, as it would permit the majority of these wastes composed to be used for commercial reprocessing, without rapidly saturating any market of a product (Wang 2010; Ucar and Wang 2011). Since the past experience has revealed, it cannot be cost-effectively viable if just a fraction of the waste carpets composed can be reused, although the rest has to be sent back to landfill. Several methods are existing, and some of them are being industrialised to reprocess fibrous wastes from discarded carpets. They vary in product superiority, possible market capacity, processing phases and price (Wang 2010).

Generally, a waste stream of a single sort of polymers such as nylon or polypropylene is easier to reprocess into products with superior quality, as compared to those wastes comprising a combination of different materials. In the carpet industries, waste fibres must be separated as early in the manufacturing phase as possible. It resulted to collect the waste fibres separately for each type and therefore easy to recycle and used in other applications. As mentioned earlier, pre-consumer waste carpet fibres which are mostly from industries are very easy to recycle and most suitable to be used in concrete construction. The properties and performance of this kind of fibres (nylon and polypropylene) are comparatively close to those virgin fibres typically used in concrete (Wang 2006). Nevertheless, it usually is hard to collect based on polymer type for post-consumer waste fibres.

In the recycling procedures, for example, nylon depolymerisation and polymeric resin revival, it is necessary to categorise the feedstock based on the kind of fibres. In case of waste carpets, the categorisation is based on the sort of the fibres used in face yarns. Melting point pointer is a tool that can recognise most fibre types, but it usually is slow (Booij et al. 1997). Dissolution or reprecipitation method has been applied to discrete the nylon from waste carpets. The diluters used contain hydrochloric acid, aliphatic alcohol and alkylphenols (Booij et al. 1997; Wang 2010). According to the process developed by Booij et al. (1997), the waste carpets were cut into 0.5–20 cm^2 pieces. The carpet shards are then mixed with the extraction agents, such as methanol. The ratio of solution to the waste carpet was typically ranged 5–20 by weight. The extraction period of 60 min was found to be adequate to dissolve nylon 6 at a temperature of 135–140 °C and 0.2–2 MPa pressure. Other solids were filtered out, cooled down the solution, and then the nylon precipitated. The gained nylon has an almost relative viscosity (about 90%) of the nylon existing in the waste carpet, representing that no severe degradation occurred in the removal procedure.

Another technology used to separation of polymers from waste carpet is melt processing. Melt processing through extrusion transforms thermoplastic polymers into resin pellets. When different kinds of polymers are mixed, the procedure is also stated as compounding. Waste carpets usually undergo a size reduction process. Waste carpets are generally bulky and need a distinct densification procedure or need to be fed to an extruder through a mainly designed crammer-densifier feeder. These methods need less working area, are cleaner and more comfortable to operate and also deserve lower energy cost and the reduction in labour work and mainte-nance costs. The extruders can be used as a single or twin screw co-rotating design dependent on the quantity, product variety and mixing intensity preferred (Hawn 2001; Strzelecki 2004; Wang 2010). Figure 4 displays a typical post-consumer carpet recycling loop.

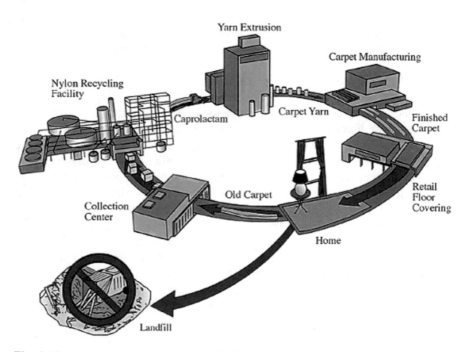

Fig. 4 Typical post-consumer carpet recycling loop

3   Concrete Composites Incorporating Waste Carpet Fibres

3.1   Fresh Properties

3.1.1   Density

The density of concrete largely depends on the unit weight of the constituent materials used in the mixture. Figure 5 displays that the unit weight of the concrete mixtures reduced with rising in fibre dosages. It can be seen that, although the fresh density was higher in plain OPC concrete as associated to those of concrete containing carpet fibres and POFA, it is lower than 2400 kg/m^3 specified by the BS EN 12350-6 (2009). It could be owing to the low density (910 kg/m^3) of the PP carpet fibres which is comparatively lower than that of conventional plain concretes. In addition, POFA-based concrete composites obtained lesser density than that of OPC mixes. This reduction in density is owing to the lower unit weight of POFA particles associated with that of OPC. Figure 5 shows that the lowest density was attained for the mix comprising 20% POFA and 1.25% fibres, which was about 4% lesser than that of the control plain concrete mix. However, the density values for those of carpet fibre and POFA are in the range of 2200–2400 kg/m^3 as regular weight concrete (BS EN 12350-6, 2009).

3.1.2   Air Content

The results of the air content test for concrete composites comprising carpet fibres are demonstrated in Fig. 6. As can be seen in the figure, the air content of concrete

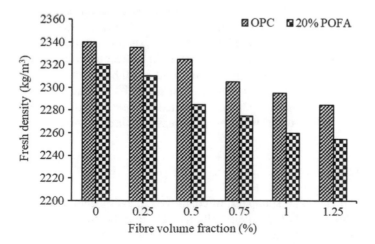

Fig. 5 Influence of carpet fibres on the fresh density of various concrete mixes

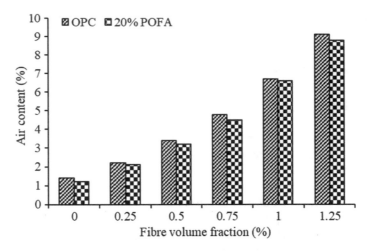

Fig. 6 Influence of carpet fibres on the air content of various concrete mixes

composites raised with increasing in carpet fibre content. For control mix, the air content was recorded as 1.4% although it quickly increased with the rise in fibre dosages. The volume of air in matrix was recorded as 2.3, 3.5, 4.7, 6.8 and 9.2% for concrete composites comprising 0.25, 0.5, 0.75, 1 and 1.25% waste carpet fibres, respectively. The higher air content in the mixes containing carpet fibres could be accredited to the minor compaction owing to the existence of fibres that provide a harsh mixture and consequently caused a higher quantity of voids and air bubbles in the fresh concrete.

Figure 6 also displays that the air content of concrete composites containing POFA is lower than concrete made with OPC alone. In fact, the large quantity of air was replaced with POFA particles, and the rate of reduction varied linearly. The values of 2.2, 3.3, 4.6, 6.8 and 8.7% were obtained for air content test of POFA-based mixes with fibres content of 0.25, 0.5, 0.75, 1 and 1.25%, respectively. The air content was found lower in all fibrous concrete mixes with POFA as related to that of fibrous OPC concrete composites. It can be owing to the increased amount of fines in the concrete mix which leans to fill both macro- and micropores inside the concrete matrix, thus causing decrease in the air voids. The outcomes of this study are in agreement with those reported by Toutanji (1999) and Akça et al. (2015) on the effects of PP fibres in an increase of air content in fresh concrete.

3.1.3 Slump

The obtained values for slump test of concrete composites comprising waste carpet fibres are shown in Fig. 7. The results indicate that the addition of waste carpet fibres to the fresh concrete composites decrease the workability. More fibre content in the matrix also revealed a lower uniformity in the mixture, which has resulted in

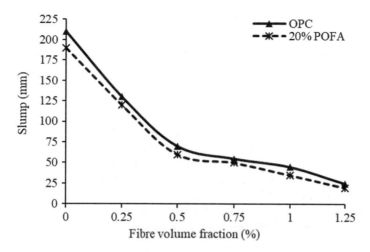

Fig. 7 Influence of carpet fibres on the slump of various concrete mixes

the balling effect of fibres and consequently dryer concrete mix, as exposed in Fig. 8. It can be observed from the results that the slump of the control plain concrete mix was recorded as 210 mm. However, by adding carpet fibres at dosages of 0.25, 0.5, 0.75, 1 and 1.25%, the slump dropped to 130, 70, 55, 45 and 25 mm, correspondingly. Owing to the existence and big surface area of polypropylene carpet fibre, extra binders paste and sands are wrapped around the fibres, and it is attributed to the strong bond of fibre–matrix in the concrete mixture which resulted in a reduction of workability (Fig. 8).

Adding POFA as cement replacement to the concrete composites generally densifies the mixture through filling up the cavities in the matrix. Consequently, it caused a harder mix with an equivalent drop in the slump values of the concrete composites. In the POFA mixtures, a rise in the fibre dosage also revealed the similar tendency in the reduction of the slump values. The values of 120, 60, 50, 35 and 20 mm were noted in POFA-based fresh mixtures for the same fibre dosages,

(a)   **(b)**   **(c)**

Fig. 8 Balling effect of fibres on fresh concrete with **a** 0.25%, **b** 0.5% and **c** 1.0% carpet fibres

correspondingly. The founding of this study is in agreement with those found by Toutanji (1999), Hsie et al. (2008) and Yap et al. (2013) that reported on the reduction of slump values by adding PP fibres to the concrete mixtures.

3.1.4 Vebe Time

Figure 9 represents the values obtained from the Vebe time test. Figure 9 reveals a linear relationship between the fibre volume fractions and the Vebe times that shows the moment of compaction to the level surface by vibration. This increasing linear trend showed that the addition of carpet fibres in the concrete mixes would lead to a reduction in workability of FRC mixtures. For plain concrete without any fibres (B1), the Vebe time was found to be 2.3 s. By adding fibres at dosages of 0.25, 0.5, 0.75, 1 and 1.25%, the Vebe time values increased to 4.5, 6.6, 8.5, 11.5 and 14.9 s, respectively.

Figure 9 shows that with the substitution of 20% POFA, the Vebe times increased more than that of OPC mixtures. For the POFA-based concrete without any fibres (B7), the Vebe time was found to be 3.6 s which was greater than that of the control mixture (B1). The time taken was increased rapidly for all fibre volume fractions; when 0.25, 0.5, 0.75, 1 and 1.25% were applied in the POFA-based fresh concrete composites, the Vebe times increased to 5, 7.1, 9.6, 12.7 and 15.5 s, respectively. This shows that all mixtures containing carpet fibres and POFA have poor workability as compared to those of OPC concrete mixtures. Though with PP fibres, a similar observation has been reported by Bentur and Mindess (2007) and Nili and Afroughsabet (2010) for FRC with relatively 70% lower workability by the addition of 0.5% PP fibres.

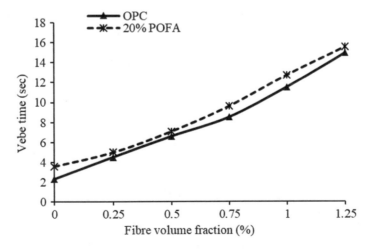

Fig. 9 Influence of carpet fibres on the Vebe time of various concrete mixes

3.2 Hardened Properties

3.2.1 Compressive Strength

Figure 10 shows that the inclusion of carpet fibres in plain concrete resulted in a reduction of compressive strength by 6, 7.5, 11.15, 18.06 and 21.23% for 0.25, 0.5, 0.75, 1.0 and 1.25% carpet fibre content, correspondingly, at the curing period of 28 days. This reduction could be owing to the low MOE of PP fibres which is 3.5–4.9 GPa and categorised this fibre as a soft material. Addition of carpet fibres to the concrete mixture will consider the concrete as a soft composite and, consequently, caused lesser compressive strength values than that of conventional concrete.

The compressive strength of concrete composite containing 0.5% carpet fibre and 20% POFA decreased by 18.3, 16.5 and 5.3% as related to that of OPC-based mixtures with the similar fibres content at the ages of 7, 28 and 91 days, correspondingly. This fall in strength values of the POFA-based composites is mainly

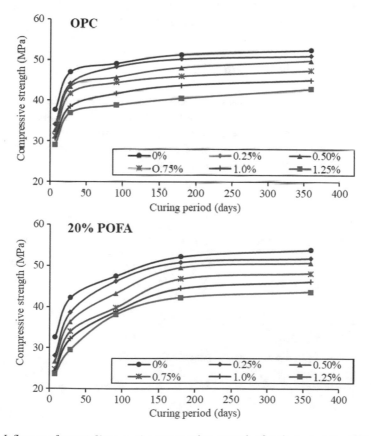

Fig. 10 Influence of carpet fibres on the compressive strength of various concrete mixes

owing to the low hydration action of POFA, which prevent the rise in strength of concrete. Lower creation of C-S-H gel in POFA concrete mixtures might have also affected the compressive strength. However, at 180 days, POFA-based concrete composites attained greater strengths than those of OPC mixes. A similar tendency was also observed for POFA mixes at the age of 365 days. The obtained results of this study are comparable to that stated by Song et al. (2005) and Nili and Afroughsabet (2010) on the reduction of compressive strength with the inclusion of PP fibres in concrete composites.

Generally, the properties of fibres control the strength properties of fibre-reinforced concrete. Fibres with greater tensile strengths can transfer more stresses from cracks to the fibres (Yap et al. 2013). The reduction in the compressive strengths of concrete composites could be owing to the existence of cavities and pores which formed while compacting process due to the lower workability of concrete (Karahan and Atis 2011). The observations made and obtained results of this study indicated that the decrease in compressive strengths of concrete composites is not noteworthy with the addition of carpet fibres. It has also been found that the failure mode of concrete composites containing carpet fibres under compression load was ductile as illustrated in Fig. 11.

3.2.2 Splitting Tensile Strength

Figure 12 indicates that the tensile strength values of concrete composites comprising waste carpet fibres were considerably more significant than that of the conventional plain concrete. It can be seen that the mixture of carpet fibre and POFA contributed to the enhancement of tensile strength of concrete composites. While the splitting happened, these fibres link the split portions of the concrete

Fig. 11 Failure mode of concrete composites with and without carpet fibres under compression load

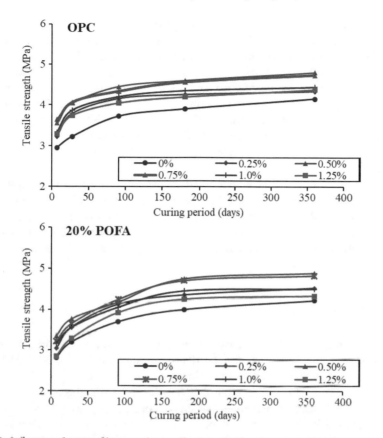

Fig. 12 Influence of carpet fibres on the tensile strength of various concrete mixes

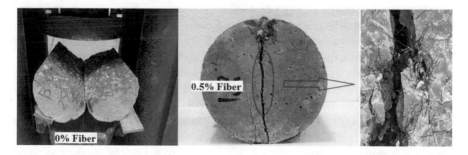

Fig. 13 Failure mode of concrete specimens without and with carpet fibres under tensile load

samples through their bridging action and transfer the stresses from the matrix to the fibre and consequently gently sustained the full tensile stresses. These transferred stresses improved the tensile strain capacity and therefore enhanced the tensile strength of concrete composites over the plain concrete counterpart as shown in Fig. 13.

The tensile strength values of OPC composites at 28-day curing improved by 17.5, 25.6, 26.2, 20.3 and 16.2% for the carpet fibre content of 0.25, 0.5, 0.75, 1 and 1.25%, correspondingly, as related to that of conventional plain concrete. The results show that the substitution of POFA in the fibrous mixes significantly contributed to the improvement of concrete tensile strength values. Nevertheless, the rate of improvement was lower particularly at the early ages owing to the low pozzolanic activity of POFA. The tensile strength of POFA concrete composites increased by 10.7, 12.5, 14.2, 8.8 and 5.5% for the same fibre content at 91-day curing period. All POFA-based composites obtained greater tensile strengths than that of OPC mix at both curing ages of 180 and 365 days. The development in the tensile strength values could be attributed to the stronger contact amongst carpet fibres and the paste–aggregate matrix as well as higher hydration products, which are in turn owing to the pozzolanic nature of POFA that grows with curing periods. The result of a present study supported with Hsie et al. (2008) and Yap et al. (2013) that reported on the enhancement of tensile strength of concrete by adding short PP fibres.

3.2.3 Flexural Strength

The flexural strength values of concrete composites were recorded in the range of 3.3–4.4 MPa, 4.2–5.5 MPa and 5–6.1 MPa for curing periods of 7, 28 and 91 days, correspondingly, with rising the carpet fibre dosages from 0 to 1.25%. The obtained results as shown in Fig. 14 indicate that the flexural strengths of concrete composites having waste carpet fibres enhanced considerably as related to that of control mix. For instance, at the age of 91 day, the strength of 6.1 MPa was found for the composite mix having 0.5% carpet fibres, which is about 20% higher than that of the plain concrete mix. A similar tendency like that of tensile strength was also observed in flexural strength of POFA-based mixes. For instance, at 91-day curing, the flexural strength of POFA-based concrete composite containing 0.5% carpet fibres was found to be 14% higher than that of plain POFA concrete mixture. Similar tendency like that for compressive and tensile strengths occurred for flexural strength at the ultimate ages. At curing periods of 180 and 365 days, all concrete mixes comprising 20% POFA performed greater flexural strengths than that of OPC concrete mixes.

As mentioned earlier, the higher values of flexural strength can be owed to the stronger contact amongst carpet fibre and the paste matrix resultant from the higher volume of hydrated products such as C-S-H gel, due to the pozzolanic behaviour of POFA, mainly at the ultimate ages. The enhancement in strength properties of concrete related to the curing time and also the pozzolanic action of POFA will be discussed with the support of microstructural analysis in the subsequent sections in this chapter. The result of the present study is in agreement with those stated by Alhozaimy et al. (1996), Martínez-Barrera et al. (2005) and Yin et al. (2016). They stated that the mixture of pozzolanic ashes and PP short fibres increased the flexural strength up to 30%.

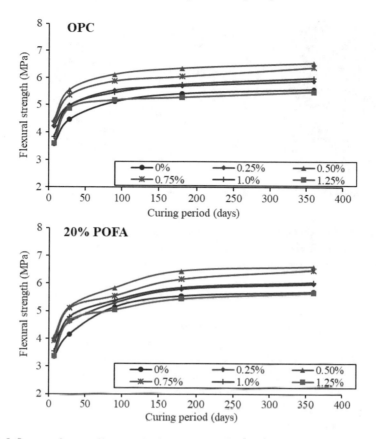

Fig. 14 Influence of carpet fibres on the flexural strength of various concrete mixes

The development of flexural strength of concrete composites is mostly due to the bridging action of carpet fibres, and also, fibres act as crack arrester in the tension zone of the concrete components. Also, carpet fibres increased the energy absorption capacity of the concrete composites and enhanced the brittleness nature of concrete through delaying in fracture and failure of concrete specimens under loads as illustrated in Fig. 15. Nevertheless, fibre dosage higher than 0.75% caused the fall of the flexural strengths. This reduction could be due to the lower workability of concrete mixtures at high dosages of fibres which disturbed the compaction and therefore formed pores that consequently resulted in lower strength in concrete (Mastali et al. 2016).

3.2.4 Impact Resistance

The impact resistance of concrete composites comprising waste carpet fibres was investigated by drop weight hammer test by the number of hammer blows required

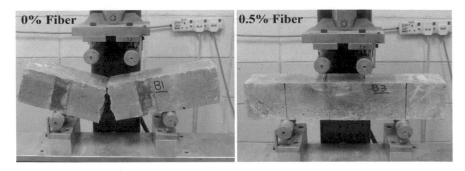

Fig. 15 Failure mode of concrete specimens without and with carpet fibres under flexure load

to attain the first crack ($N1$) and the failure ($N2$) of the specimens. The obtained results given in Table 1 reveal that by adding waste carpet fibres in the concrete mixture, the impact resistance of concrete significantly enhanced. It can be observed that the number of blows recorded for mixes containing 0.025, 0.5, 0.75, 1.0 and 1.25% was increased by 54, 158, 221, 300 and 367%, respectively, as compared to plain concrete mix. The higher number of blows for mixes containing carpet fibres indicates that the fibres considerably improved the ductility of the concrete composites. The results further show that the substitution of POFA in concrete composites caused a slight reduction in the impact resistance of mixes. This might be due to the lower strength development of POFA-based mixes due to the slow pozzolanic reaction of POFA mainly at early times. Nevertheless, the combination of carpet fibre and POFA showed an adequate impact resistance of concrete composites and higher energy absorption as related to that of conventional plain concrete.

Table 1 Impact resistance of concrete composites at first crack and failure

Mix	Impact resistance		$N2 - N1$	Impact energy (kN mm)	
	First crack ($N1$)	Failure ($N2$)		First crack	Failure
B1	24	28	4	488.6	569.7
B2	37	48	11	753.2	977.2
B3	62	81	19	1261.8	1648.6
B4	77	95	18	1567.4	1933.5
B5	96	117	21	1954.1	2381.6
B6	112	145	33	2279.5	2951.4
B7	19	22	3	386.6	447.9
B8	32	44	12	651.5	895.8
B9	55	73	18	1119.4	1485.6
B10	69	85	16	1404.5	1730.3
B11	90	111	21	1831.7	2259.5
B12	105	138	33	2137.3	2808.7

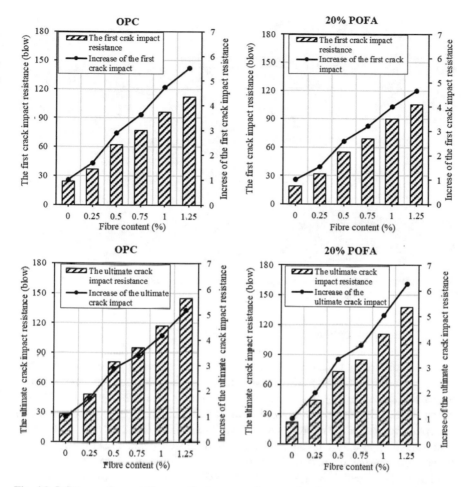

Fig. 16 Influence of carpet fibres on the impact resistance of various concrete composites

Based on the obtained results and observation made, the mixture of carpet fibres and POFA considerably improved the ductility performance and post-peak resistance of the concrete composites (Fig. 16). Figure 17a, b shows the concrete samples used for the drop weight test and the crack patterns after the fracture. Increasing carpet fibre content caused higher cracks on the top surface of specimens, as exposed in Fig. 17c, d. The cracks developed on the concrete specimens owed to the linking action of carpet fibres which increase the energy absorption capacity and banned the quick failure of the specimens. A fractured surface is also illustrated in Fig. 17e, where the portions of the waste carpet fibres are visible. According to the given fractured surface, it is exposed that carpet fibres are uniformly distributed along the section.

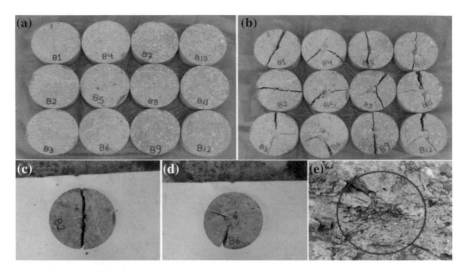

Fig. 17 **a** Samples before loading, **b** crack patterns after drop weight test, **c** 0.25% fibres, **d** 1.25% fibres and **e** dispersed carpet fibres

3.2.5 Water Absorption

The permeability of concrete composites in terms of water absorption test was investigated for OPC and POFA mixes including carpet fibres, and the obtained results are shown in Fig. 18. As expected, concrete permeability decreased with curing period. The results reveal that by the addition of waste carpet fibres the water absorption of concrete composites significantly reduced. It can be seen that for the fibre dosage up to 0.5%, the water absorption dropped by about 4%. However, rise in the fibre dosage beyond 0.5% caused more absorption of water in concrete specimens. This greater absorption of water could be owing to the existence of large amount of fibres and balling effect caused by fibres, and therefore, lower compaction resulted in the occurrence of voids. Additionally, the bridging action of carpet fibres significantly reduced the volume of microcracks in the matrix, which caused lesser penetration of water into the concrete specimens. The attained outcomes are similar to those found by Seleem et al. (2010) and Liu et al. (2014). They reported that the inclusion of short PP fibres is potential to reduce the water absorption of concrete specimens.

From Fig. 18, it also can be observed that OPC concrete mixes absorbed less water at 28 days of curing as related to the specimen containing 20% POFA. Concrete containing POFA absorbs less water than the control specimen when cured for a period longer than 28 days. At 90-day curing, the water absorption of all fibrous mixes comprising POFA was slightly lesser than that of OPC mixes. The relative decrease in the absorption capacities is probably due to the increase in the paste volume POFA and fineness of the ash, which occupied both the macro- and micropores in the mix. In other words, the POFA content has influenced pore and

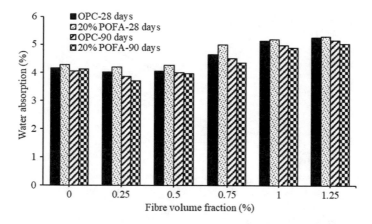

Fig. 18 Water absorption versus fibre volume fraction of OPC and POFA concretes

grain refinements of the concrete. Although not with POFA, a comparable observation was made by Karahan and Atis (2011), who reported related findings on the influence of fly ash and PP fibres on the reduction of water absorption of concrete.

3.2.6 Chloride Penetration

The depth chloride penetration test of the concrete composites was investigated through the immersion of concrete specimens in 5% NaCl solution. Figure 19 displays the experimental outcomes of the chloride penetration test at the ages of 7, 28 and 90 days. It can be observed that the addition of carpet fibres up to a particular dosage reduced the depth of chloride penetration for all curing periods. The uniform distribution of carpet fibres in the concrete mix was able to provide a network structure, which had a constructive influence in the reduction of pores and prevents the entrance of disturbing particles into the specimens and, therefore, resulted in a lower depth of chloride penetration.

The depth of diffusion for both OPC and POFA composites having 0.25, 0.5 and 0.75% waste carpet fibre was noticeably reduced. It can be observed that the depth of chloride diffusion in the OPC mix with 0.5% fibre was found as 14 mm, which is about 27% slighter than that of 19 mm found for the plain concrete mixture. A comparable propensity was perceived in the POFA-based concretes. The concrete composites comprising 0.5% fibre exhibited 10 mm penetration depth at 90-day curing. This reduction in the depth of penetration is due to the addition of carpet fibres which were uniformly distributed in the concrete matrix and provide a grid structure which prevents the entrance of any disturbance particle into the concrete. Nevertheless, by the addition of fibres at higher dosages, the depth of penetration was gradually increased by increasing the permeability of concrete specimens. Moreover, by the replacement of POFA particularly at the ultimate ages, the depth

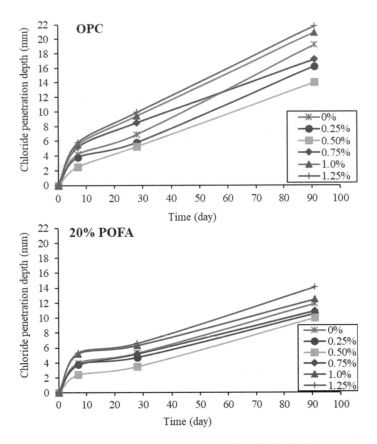

Fig. 19 Influence of carpet fibres on the chloride penetration depth of various concrete mixes

of penetration reduced. It could be attributed to the finer size of the POFA particles which filled the pores through its pozzolanic reactions such as extra C-S-H gels (Chindaprasirt et al. 2008).

3.3 Microstructural Analysis

The scanning electron micrograph analysis for concrete mixes at the ages of 91 and 365 days is exposed in Fig. 20. Invariably, the microstructure of the respective concrete reformed as the curing time progresses. As shown in Fig. 20a, at 91 days of curing period, the hydration products such as C-S-H gels are more steadily spared in a matrix with POFA than that of OPC mix. It indicates that OPC mixture comprises numerous crystalline which interweaves together with C-S-H gel with some cavities are visible amongst these crystalline. While POFA-based mix comprises several gel constituents and has much fewer pores than OPC-based mix.

(a)

(b)

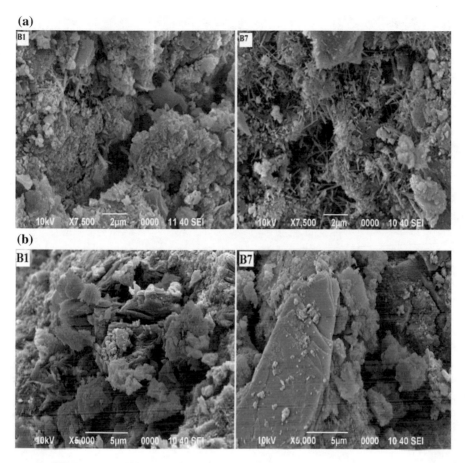

Fig. 20 SEM of OPC (B1)- and POFA (B7)-based concrete specimens at **a** 91 and **b** 365 days

Figure 20b further illustrates the uniform distribution of C-S-H gels and the development of the new C-S-H gels at the age of 365 days. The better performance of POFA-based mixture is due to the intake of portlandite during the pozzolanic reaction of POFA at the ultimate ages. At 365-day curing, POFA concrete composites attained greater strengths as related to those of OPC composites. At this age, the hydration products of the POFA are well developed. The structure of the POFA paste comprises POFA particle in the state of more advanced hydration. This demonstrates that CH crystals have been consumed and transformed into C-S-H, which considered the strength is contributing phase. A similar observation was reported by Uygunoglu (2008), Şahmaran et al. (2011) and Ranjbar et al. (2016) who reported the enhancement in the microstructure of concrete composites incorporating PP fibres and pozzolanic materials.

Figure 21 displays the SEM of fibre–matrix interface of the matrix comprising 0.5% carpet fibres and also the bridging action of fibres. The outcomes of the

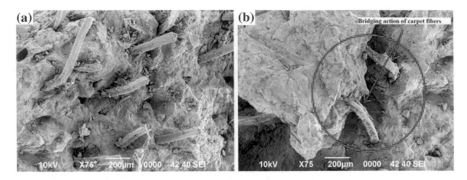

Fig. 21 SEM images of **a** uniformly spared fibres and **b** bridging action of carpet fibres in concrete composites

flexural and tensile strengths and impact resistance of concrete composites comprising waste carpet fibre signify a good fibre–matrix interface and stronger fibre–matrix bond. Furthermore, while mixing, fibre monofilaments remain anchored to the fibres and cement paste. In a short period during the hydration process, it is significant to wait for hydration products for increasing the bond.

Figure 22 shows the fibre–cement interfacial transition zone and also reflects a robust bond amongst fibre and cement paste cured water. The SEM image also shows that the fibre along with cement paste provided a robust interfacial bonding that results in the decrease of the crack size. Therefore, a decrease in the cracks increases the strength and energy absorption. Enhancement in ductility of concrete composites comprising waste carpet fibre is owing to the linking action of fibres, through which these fibres can moderately transfer the stress across the crack.

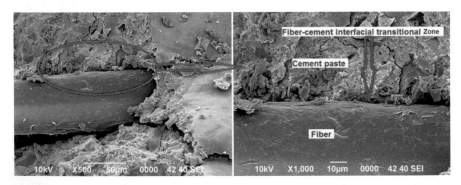

Fig. 22 SEM of the fibre–cement interfacial transition zone

4 Applications

Concrete composites are more and more used on account of the benefits of higher tensile and flexural strengths, more significant energy absorption capacity and the better ductility performance. The consistent distribution of short fibres to the concrete mixture offers isotropic properties not common to conventional steel-reinforced concrete. Since its introduction to the market in the late 1960s, the use of fibre-reinforced concrete has increased steadily. The following are the most common applications that concrete composites incorporating carpet fibres can be used:

- Concrete composites containing waste carpet fibres can be efficiently used as concrete pavements of highways and airport runways, bridges components, housing and industrial floors and other similar applications.
- This type of concrete composites can be used as utility concrete components such as underground vaults and junction boxes, drains for industrial wastes, sewer channels and pipes and power line transmission poles.
- Concrete composites comprising carpet fibres are potential to be used as railroad ties, median barriers and bridge panels in transportation-related works.
- The existing sorts of polypropylene industrial waste carpet can be used to fabricate marine construction materials that are economically competitive and environmentally superior to conventional marine construction products.

5 Conclusions

- The presence of carpet fibre in concrete decreased the unit weight of the mixture. The higher the fibre dosage was, the lower the unit weight due to the lower density (910 kg/m^3) of carpet fibres compared to that of ordinary concrete. Moreover, the volume of air in concrete was increased by the addition of carpet fibres.
- The inclusion and further rise in carpet fibre dosage resulted in lower workability of concrete composites. Generally, higher fibre content leads to lower slump and higher Vebe time.
- The addition of waste carpet fibres resulted in lower compressive strength values of concrete composites at early ages. However, for the POFA-based composites, the compressive strength enhanced at the ultimate curing periods.
- Adding carpet fibres to mixtures significantly enhanced the splitting tensile strength and the flexural strength of the concrete composites. Replacement of POFA together with carpet fibres caused superior strength values mostly at the ultimate ages owing to the bridging action of carpet fibre and pozzolanic nature of POFA.

- The impact resistance and the energy absorption capacity of concrete composites significantly increased by adding carpet fibres.
- The mixture of carpet fibre and POFA expressively improved the permeability of concrete composites and therefore resulted in lower water absorption. The influence of POFA on the reduction of water absorption was more noticeable at more extended curing periods.
- Adding carpet fibres to the POFA-based concrete composites up to 1% significantly reduced the depth of chloride penetration.
- The SEM results revealed sturdy fibre–cement interfacial transition zones and also reflect a strong bond between carpet fibre and binders paste cured in water. Therefore, drop in the number of cracks increases the strength and energy absorption.

The outcomes of this chapter recommend that concrete composite comprising industrial waste carpet fibres and palm oil fuel ash is potential to manufacture and is industrialised with adequate engineering properties and durability performance. Nevertheless, to ensure the feasibility of this new type of concrete composite as structural components, large-scale applications together with its behaviour in steel-reinforced concrete components are suggested for the upcoming studies.

References

Akça KR, Çakır Ö, Ipek M (2015) Properties of fiber reinforced concrete using recycled aggregates. Constr Build Mater 98:620–630

Aldahdooh MAA, Bunnori NM, Johari MAM (2014) Influence of palm oil fuel ash on ultimate flexural and uniaxial tensile strength of green ultra-high performance fiber reinforced cementitious composites. Mater Des 54:694–701

Alhozaimy A, Soroushiad P, Mirza F (1996) Mechanical properties of polypropylene fiber reinforced concrete and the effects of pozzolanic materials. Cement Concr Compos 18:85–92

Al-mulali MZ, Awang H, Khalil HA, Aljoumaily ZS (2015) The incorporation of oil palm ash in concrete as a means of recycling: a review. Cement Concr Compos 55:129–138

Awal ASMA, Hussin MW (1997) The effectiveness of palm oil fuel ash in preventing expansion due to alkali-silica reaction. Cement Concr Compos 19:367–372

Awal ASMA, Mohammadhosseini H (2016) Green concrete production incorporating waste carpet fiber and palm oil fuel ash. J Clean Prod 137:157–166

Batayneh M, Marie I, Asi I (2007) Use of selected waste materials in concrete mixes. Waste Manag 27(12):1870–1876

Bentur A, Mindess S (2007) Fibre reinforced cementitious composites. Taylor & Francis, London

Bird L (2013) Carpet recycling UK. In: Carpet recycling UK conference, London

Booij M, Hendrix JAJ, Frentzen YH (1997) Process for recycling polyamide-containing carpet waste. Eur Pat 759:456

Brandt AM (2008) Fibre reinforced cement-based (FRC) composites after over 40 years of development in building and civil engineering. Compos Struct 86(1):3–9

Chandra S (1997) Waste materials used in concrete manufacturing. Noyes Publications, New Jersey

Chindaprasirt P, Rukzon S, Sirivivatnanon V (2008) Resistance to chloride penetration of blended Portland cement mortar containing palm oil fuel ash, rice husk ash and fly ash. Constr Build Mater 22(5):932–938

Ghosni N, Samali B, Vessalas K (2013) Evaluation of mechanical properties of carpet fibre reinforced concrete. From materials to structures: advancement through innovation. Taylor and Francis Group, London, pp 275–279

Gu L, Ozbakkaloglu T (2016) Use of recycled plastics in concrete: a critical review. Waste Manag 51:19–42

Hawn KL (2001) An overview of commercial recycling technologies and textile applications for the products. In: Presentation at 6th annual conference on recycling of polymer, Textile and Carpet Waste, Dalton

Hsie M, Tu C, Song PS (2008) Mechanical properties of polypropylene hybrid fiber-reinforced concrete. Mater Sci Eng, A 494(1–2):153–157

Ismail M, Ismail ME, Muhammad B (2011) Influence of elevated temperatures on physical and compressive strength properties of concrete containing palm oil fuel ash. Constr Build Mater 25(5):2358–2364

Kanadasan J, Abdul Razak H (2015) Engineering and sustainability performance of self-compacting palm oil mill incinerated waste concrete. J Clean Prod 89:78–86

Karahan O, Atiş CD (2011) The durability properties of polypropylene fiber reinforced fly ash concrete. Mater Des 32(2):1044–1049

Liu J, Xing F, Dong B, Ma H, Pan D (2014) Study on water sorptivity of the surface layer of concrete. Mater Struct 47(11):1941–1951

Martínez-Barrera G, Vigueras-Santiago E, Hernández-López S, Brostow W, Menchaca-Campos C (2005) Mechanical improvement of concrete by irradiated polypropylene fibers. Polym Eng Sci 45(10):1426–1431

Mastali M, Dalvand A, Sattarifard A (2016) The impact resistance and mechanical properties of self-compacting concrete reinforced with recycled CFRP pieces. J Clean Prod 124:312–324

Meddah MS, Bencheikh M (2009) Properties of concrete reinforced with different kinds of industrial waste fibre materials. Constr Build Mater 23(10):3196–3205

Mo KH, Alengaram UJ, Jumaat MZ, Liu MYJ (2015) Contribution of acrylic fibre addition and ground granulated blast furnace slag on the properties of lightweight concrete. Constr Build Mater 95:686–695

Mohammadhosseini H, Awal ASMA (2013) Physical and mechanical properties of concrete containing fibers from industrial carpet waste. Int J Res Eng Technol 2(12):464–468

Mohammadhosseini H, Yatim JM (2017) Microstructure and residual properties of green concrete composites incorporating waste carpet fibers and palm oil fuel ash at elevated temperatures. J Clean Prod 144:8–21

Mohammadhosseini H, Awal ASMA, Sam ARM (2016) Mechanical and thermal properties of prepacked aggregate concrete incorporating palm oil fuel ash. Sādhanā 41(10):1235–1244

Mohammadhosseini H, Yatim JM, Sam ARM, Awal ASMA (2017) Durability performance of green concrete composites containing waste carpet fibers and palm oil fuel ash. J Clean Prod 144:448–458

Mohammadhosseini H, Tahir MM, Sam ARM, Lim NHAS, Samadi M (2018) Enhanced performance for aggressive environments of green concrete composites reinforced with waste carpet fibers and palm oil fuel ash. J Clean Prod 185:252–265

Nili M, Afroughsabet V (2010) The effects of silica fume and polypropylene fibers on the impact resistance and mechanical properties of concrete. Constr Build Mater 24:927–933

Putman BJ, Amirkhanian SN (2004) Utilization of waste fibers in stone matrix asphalt mixtures. Resour Conserv Recycl 42(3):265–274

Ranjbar N, Behnia A, Alsubari B, Birgani PM, Jumaat MZ (2016) Durability and mechanical properties of self-compacting concrete incorporating palm oil fuel ash. J Clean Prod 112:723–730

Şahmaran M, Özbay E, Yücel HE, Lachemi M, Li VC (2011) Effect of fly ash and PVA fiber on microstructural damage and residual properties of engineered cementitious composites exposed to high temperatures. J Mater Civ Eng 23(12):1735–1745

Schmidt H, Cieslak M (2008) Concrete with carpet recyclates: suitability assessment by surface energy evaluation. Waste Manag 28:1182–1187

Seleem HEDH, Rashad AM, El-Sabbagh BA (2010) Durability and strength evaluation of high-performance concrete in marine structures. Constr Build Mater 24(6):878–884

Siddique R, Khatib J, Kaur I (2008) Use of recycled plastic in concrete: a review. Waste Manag 28(10):1835–1852

Song PS, Hwang S, Sheu BC (2005) Strength properties of nylon- and polypropylene-fiber-reinforced concretes. Cem Concr Res 35(8):1546–1550

Sotayo A, Green S, Turvey G (2015) Environmental technology & innovation carpet recycling: a review of recycled carpets for structural composites. Environ Technol Innov 3:97–107

Strzelecki C (2004) Modern solutions for shredding, grinding and repelletizing post-industrial fiber, nonwovens and carpet scrap. In: Presentation at 9th annual conference on recycling of polymer, Textile and Carpet Waste, Dalton

Tangchirapat W, Saeting T, Jaturapitakkul C, Kiattikomol K, Siripanichgorn A (2007) Use of waste ash from palm oil industry in concrete. Waste Manag 27(1):81–88

Tay JH (1990) Ash from oil-palm waste as a concrete material. J Mater Civ Eng 2(2):94–105

Thomas BS, Gupta RC (2013) Mechanical properties and durability characteristics of concrete containing solid waste materials. J Clean Prod 48:1–6

Toutanji HA (1999) Properties of polypropylene fiber reinforced silica fume expansive-cement concrete. Constr Build Mater 13(4):171–177

Ucar M, Wang Y (2011) Utilization of recycled post-consumer carpet waste fibers as reinforcement in lightweight cementitious composites. Int J Clothing Sci Technol 23(4):242–248

Uygunoglu T (2008) Investigation of microstructure and flexural behavior of steel-fiber reinforced concrete. Mater Struct 41:1441–1449

Wang Y (2006) Utilization of recycled carpet waste fibers for reinforcement of concrete and soil. Recycling in textiles. Woodhead Publishing Ltd., Cambridge, pp 1–14

Wang Y (2010) Fiber and textile waste utilization. Waste Biomass Valorization 1(1):135–143

Wang Y, Zureick AH, Cho BS, Scott DE (1994) Properties of fibre reinforced concrete using recycled fibres from carpet industrial waste. J Mater Sci 29(16):4191–4199

Wang Y, Wu HC, Li VC (2000) Concrete reinforcement with recycled fibers. J Mater Civ Eng 12(4):314–319

Yahaghi J, Muda ZC, Beddu SB (2016) Impact resistance of oil palm shells concrete reinforced with polypropylene fibre. Constr Build Mater 123:394–403

Yap SP, Alengaram UJ, Jumaat MZ (2013) Enhancement of mechanical properties in polypropylene- and nylon-fibre reinforced oil palm shell concrete. Mater Des 49:1034–1041

Yap SP, Bu CH, Alengaram UJ, Mo KH, Jumaat MZ (2014) Flexural toughness characteristics of steel–polypropylene hybrid fibre-reinforced oil palm shell concrete. Mater Des 57:652–659

Yin S, Tuladhar R, Sheehan M, Combe M, Collister T (2016) A life cycle assessment of recycled polypropylene fibre in concrete footpaths. J Clean Prod 112:2231–2242

Zhang P, Li Q, Sun Z (2011) Influence of silica fume and polypropylene fiber on fracture properties of concrete containing fly ash. J Reinf Plast Compos: 0731684411431358

Zollo RF (1997) Fiber-reinforced concrete: an overview after 30 years of development. Cement Concr Compos 19(2):107–122

Environmentally Benign and Sustainable Green Composites: Current Developments and Challenges

A. M. Mhatre, A. S. M. Raja, Sujata Saxena and P. G. Patil

Abstract Increasing human population, environmental impacts of synthetic polymers and limited non-renewable resources have steered the development of sustainable alternatives. Synthetic polymer composites are made from two or more non-biodegradable components which make its recycling difficult leading to high disposal cost and negative environmental impact. Green composites are considered an alternative to petroleum-based polymer composites due to their lower environmental impacts. Voluminous researchers have proposed the development of biodegradable green composites from renewable sources like natural fibres, agricultural feedstocks and biopolymers to make them sustainable and environmentally friendly. Furthermore, these green composites mainly constituted of natural fibres and biopolymers have high specific strength, noteworthy processing benefits, and low relative density and biodegradable which makes them industrially applicable. Although green composites are sustainable and environmentally friendly, they still need to overcome challenges like low ductility, low dimensional stability, less long-term stability for outdoor applications and moisture absorption. Even though there is scope for research in future for green composites to completely substitute synthetic polymer composites, green composites are already being used in applications which require less mechanical resistance like packaging, gardening items, automotive panels, furniture. Green composites have come a long way in the past 10 years and are expected to generate properties comparable/superior to synthetic polymer composites in near future which will play a remarkable role in reducing usage of non-renewables and waste generation.

Keywords Green composites · Sustainability · Challenges · Natural fibres Biopolymers · Industrial applicability

A. M. Mhatre · A. S. M. Raja (✉) · S. Saxena · P. G. Patil
ICAR-Central Institute for Research on Cotton Technology, Adenwala Road, Matunga, Mumbai 400019, India
e-mail: asmraja16475@gmail.com

S. S. Muthu (ed.), *Green Composites*, Textile Science and Clothing Technology, https://doi.org/10.1007/978-981-13-1969-3_3

1 Introduction

In the world of increasing global temperatures and irreversible impact of synthetic materials on ecosystems, tenacious efforts are being made to develop environment-friendly green technologies to reduce the environmental impact of processes and products. Disposal of materials used in different industries and products is essential as waste accumulation affects aquatic life and has dumping issues. Around 90% of composites end up in landfills, and only 10% is recycled or incinerated. According to Eriksen et al. (2014), 5.25 trillion plastic particles weighing 268,940 tons are currently floating at sea. Development of biodegradable materials has been a topic of interest for researchers since more than a decade, and different biopolymers have been developed and being used in different industries. Although biopolymers possess properties comparable to synthetic materials, they still cannot substitute synthetic polymers in all applications because of their high cost, less shelf life and processability. Overall properties of polymers can be improved by reinforcing them on a polymeric matrix, and these reinforced polymers are called composites. Composites contain matrix material, reinforcing material and interface which keeps them both together. Matrix material can be polymer, metal, ceramic and carbon, and reinforcing materials are fibres, whiskers, microspheres, flake, particulates, etc. Different methods of making composites are mainly filament winding, layup methods, resin transfer moulding, injection moulding and autoclave bonding (Hills 2011). Composites exhibit better mechanical and physical properties compared to polymers. Composites can be divided into three types: non-biodegradable composites, partially biodegradable composites and biodegradable composites. Non-biodegradable composites contain petroleum-derived synthetic polymers; matrix and reinforcing materials are non-biodegradable. Partially degradable composites have either matrix material or reinforcing material of non-biodegradable nature, for example sisal fibre reinforced on unsaturated polyester and polypropylene (Bajpai et al. 2014). Disposal of partially degradable composites is difficult as it cannot be recycled or completely degraded. Biodegradable composites contain biodegradable matrix and biodegradable reinforcement, for example sisal fibre reinforced on rubber seed oil polyurethane resin (Bajpai et al. 2014). Green composites mainly contain natural fibres and biopolymers. Natural fibres are of two types: vegetable or animal fibres. Vegetable fibres mainly comprise cellulose and hemicelluloses, whereas animal fibres contain proteins. Biopolymers are chiefly derived from renewable sources which include thermoplastics like polylactic acid, polyhydroxybutyrates, plant and animal sources like cellulose, chitin. Green composites which contain natural fibres and biopolymers are completely renewable, and green composites which contain one of the components as synthetic polymer and other component as biodegradable are partially biodegradable. Partially degradable green composites cannot be called environmentally benign as although fillers get degraded matrix remains in the environment. Unlike composites with synthetic components which can be recycled, partially degradable green composites cannot be recycled, and hence, the only method of their disposal is incineration. Different pollutants emitted from incinerators are metals, acid gases, particulate matter, oxides of nitrogen and sulphur which lead to smog formation, eutrophication, global warming, acidification and human and animal toxicity (Sharma et al. 2013). Holistic approach is essential to develop environment-friendly green composites as biodegradable green composites do not ensure to be environmentally benign. Sources of green composite constituents are essential. Natural fibres used in green composites like jute, hemp, flax, sisal need fertile land for cultivation. Global population will increase to 9.1 billion in 2050, and to provide sufficient food to the world in 2050, 70% increase in food production will be essential (FAO 2009). It will be vital to develop products which do not need arable land for their production as fertile land will be needed for food production. Hence, composites made from biopolymers obtained from fermentation processes, biodegradable polymers and natural fibres from agricultural and industrial wastes will be a sustainable option. Developing entirely biodegradable and renewable green composites with properties comparable to conventional composites will be an essential need of future.

The chapter begins by briefly discussing different matrix materials and reinforcements used in composites and further stating categories of composites and their respective advantages and shortcomings. Furthermore, the environmental impact of various composite materials and composites as a whole is discussed briefly and composites made from environment friendly materials are introduced. Green composites which are environmentally benign are concisely discussed affirming different matrix and reinforcing materials which are biodegradable and are easily disposed of. Also, processing aspects of green composites are explored along with processing problems involved with green composite constituents. Finally, different attributes of green composites exhibit prospects and stumbling blocks in making green composites commercial and hence concluding that green composites will substitute conventional composites in many more applications in future.

2 Composites

Composites are materials which contain two or more chemically and physically distinct phases separated by an interface. Composites are different from blends as composites have recognizable phases even after reinforcement. Matrix and reinforcing materials are together because of the interface, but they retain their original properties because these composites have distinct properties like high strength and lightweight (La Mantia and Morreale 2011). Matrix phase is comparatively more ductile and holds reinforcing phase and distributes the load of reinforcing phase. Reinforcing phase is the discontinuous phase which is harder than matrix phase. Nature displays some fascinating examples of composites like wood wherein cellulose fibres are reinforced in lignin matrix (Jose et al. 2012). Polymers are common matrix materials used wherein epoxy, polyesters and vinyl esters are mostly used. Reinforcing materials are supposed to be hard and provide strength to the matrix.

Commonly used reinforcing materials are natural fibres like cellulose, hemp, jute and high strength polymers like aramid, glass fibres and carbon fibres.

2.1 Matrix Materials

A matrix material can be polymeric, ceramic and metallic, and it binds to reinforcing material, determines the quality of the surface, supports and distributes the load of reinforcing the material, and also gives shape to composite. Selection of the matrix depends on the environment in which composite is to be used. The polymer matrix is widely used in commercial applications. Metal, carbon and ceramic matrices are used in high-temperature environments like engines.

2.1.1 Polymer Matrix

The polymer matrix is different types of thermoset, thermoplastic resins and rubber. Polymers are large macromolecules formed from subunits connected with covalent bonds. They are cheaper elastomers, less dense than metals and ceramics. Polymer matrix provides low weight, atmospheric, corrosion-resistant matrix with superior electrical resistance.

Thermoset Resin Matrices

Thermosets are widely used plastic resins which when cured thermally or chemically form infusible and insoluble polymers. Unsaturated polyester resins are most widely used thermosets mainly because of their good mechanical and electrical stability along with low cost. Glass reinforcements are mainly used with unsaturated polyester resins which provide good support and stability. For advanced composite matrices, commonly used thermosets are epoxies, cyanate esters, phenolics, polyimides, bismaleimides, etc. These thermosets have unique properties and applications like epoxy resins provide high strength, durability, chemical resistance and stability at high temperatures. Cyanate resins provide high strength, superior electrical properties, and toughness with low moisture absorption. Although they have the higher cost, phenolic resins have applications in producing flame-resistant aircraft interior panels and flame-resistant low smoke products. Bismaleimide and polyimides are used in high-temperature applications in aircraft and missiles. Most of the thermosets are derived from petroleum products, and hence, there is a substantial amount of research involved in developing bioresins from plant sources. Polyols and ethanols have been developed from sources like soybean, corn to develop bioresins with comparable properties to unsaturated polyester resins.

Thermoplastic Resin Matrices

Thermoplastics unlike thermosets can be re-melted and reshaped as they have the property of plasticity which makes them reusable and recyclable which gives them the upper hand over thermosets when it comes to environmental sustainability. Thermoplastic resins like polyethylene, polyethylene terephthalate (PET), poly-butylene terephthalate, polycarbonates, acrylonitrile butadiene styrene, polypropy-lene need low processing temperature, high moisture resistance along with low cost and are used in commercial applications like footwear, orthotics, prosthetics, automotive air intake manifolds. High-performance thermoplastics perform well at high temperature and high moisture environments. Thermoplastics like polyetheretherketone, polyamideimide, polyarylsulfone, polyetherimide, poly-phenylene sulphide reinforced with high-performance fibres exhibit high shelf life and impact resistance.

2.1.2 Metal Matrix

Metal matrix provides good strength, rigidity and stiffness to composites. These matrices are stable at elevated temperatures in the corrosive environment compared to polymer composites. Reinforcement materials need to be non-reactive and stable at high temperatures to make a stable metal matrix composite. High weight of the matrix can be resolved by using metals like aluminium, titanium and magnesium which have applications in aircraft applications (Rohatgi 1993). Metal matrix provides high strength-to-weight ratio higher than most alloys.

Advantages of Metal Matrix Composites (MMC)

1. MMC have superior strength because of which MMC has a wide variety of structural applications.
2. High elastic modulus is exhibited by metal matrix composites.
3. MMC has high toughness and impact resistance.
4. MMC has high-temperature stability and is least sensitive to temperature changes or thermal shocks.
5. MMCs have low sensitivity to surface flaws and high durability.
6. Strength retention and thermal degradation at high temperatures.
7. MMCs have environmental stability at extreme conditions.

Constituents used in MMC are mainly different types of alloys; in very rare cases, single metal is used in metal matrix composites. Combinations of metals are used in MMC to give it essential mechanical and thermal properties. Different metals like aluminium, copper, iron, magnesium are often used in different MMCs depending on the cost and desired nature of the composites (Rohatgi 1993). MMCs are costly compared to polymer matrix composites because of which they have applications mainly in high value sectors (Table 1).

Table 1 Applications of metal matrix composites (Taya and Arsenault 2000)

Fibre	Matrix	Fabrication method	Application
Molybdenum	Titanium alloy	Powder metallurgy, fibre alignment by extrusion rolling, etc.	Supersonic aircraft rocket propulsion
Stainless steel	Nickel alloys	Electroforming method	Rocket engines
Niobium filaments	Nickel, copper, silver	Melt impregnation	Superconductors
Carbon	Magnesium alloys	Melt impregnation	High-pressure vessels and fan blades of turbines
Carbon-coated with silicon oxide	Aluminium, nickel, magnesium and titanium	Melt impregnation, powder metallurgy	Different engine components
Carbon	Silicon	Powder metallurgy	Abrasive materials
Aluminium oxide	Aluminium oxide	Infiltration with metal alloys	Aerospace structures, components on engines
Glass fibres	Lead	Melt impregnation of glass fibres on lead	Acoustic insulation, battery plates, bearing materials

2.1.3 Ceramic Matrix

Ceramic matrix is mainly preferred in structural applications stable at elevated temperatures as high as 1500 °C. Ceramic matrix has high surface area and covalently bond with reinforcements which make composites having high melting points, stability at elevated temperatures, good corrosion resistance and high compressive strength (Mallick 1997). Ceramic matrix is preferred for high-temperature environments. They are used in aircraft applications, applications in engines wherein high-temperature stability is essential. Ceramic materials are crystalline in nature. Ceramic materials lack the property of plasticity like metals and plastics because of which deformation and brittleness are prominent in ceramics, and hence, lack of strength is observed in ceramics (Chawla 2013). Even small crack can lead to breakage of ceramic materials. Fibre reinforcements increase the toughness of ceramics which someway hides the low toughness of ceramics. Ceramics are reinforced with discontinuous reinforcements like whiskers, particles, short fibres etc. and continuous reinforcements like long fibres, choice of reinforcements depends on desirable properties ceramic composites need to possess based on their applications.

Advantages of Ceramic Matrix Composites

1. Ceramic composites have high corrosion resistance and wear in different conditions and temperatures.
2. Ceramic composites are stable at high temperatures and retain strength at elevated temperatures.

3. Composite matrix has high strength at comparatively less increase in weight of the composite.
4. Composites are less reactive in nature and have high chemical stability.
5. Reinforcing ceramic composites with high tensile strength fibres provides them with higher hardness.
6. Ceramic composites are lighter in weight compared to MMC because of which they can be used in low weight-to-strength applications.
7. Ceramic matrix composites exhibit non-catastrophic failure.

Disadvantages of Ceramic Matrix Composites

1. Ceramic matrix composite processing needs high temperatures and can only be reinforced with high-temperature reinforcements.
2. Brittleness is one of the main disadvantages of ceramic matrix composites even though reinforcements improve their toughness, but small cracks can lead to ceramic degradation.
3. Ceramic matrix composites need high temperatures for processing which makes the manufacturing process more complex which increases the processing cost.
4. Reinforcements and ceramic matrix materials have the different thermal coefficient of expansion which leads to thermal stresses while cooling after processing step.

Applications of Ceramic Matrix Composites

Ceramic matrix composites have applications in the fields where high temperatures, wear and corrosion resistance are the prime criteria. Silicon carbide reinforcement on aluminium oxide is used in cutting tools for machining hard materials. Ceramic matrix composites also have applications in making components of gas turbines, shielding systems, engines used in aerospace sector, different components used in burners, flame holders, components of bearing which have high probability of getting corroded, braking components like brake discs and brake system components used in cars and airplanes which might be exposed to sudden high temperatures and hot gas ducts.

2.2 Reinforcing Materials

Reinforcing materials in composites can be mainly of three types based on the nature and properties of reinforcing materials. These are fibre-reinforced composites, particle-reinforced composites and structural composites.

2.2.1 Fibre-Reinforced Composites

Fibres are widely used as reinforcing materials mainly due to their high strength per unit weight. Synthetic fibres like Kevlar are widely used in applications like the

Table 2 Properties of different natural fibres (Koronis et al. 2013)

Fibres	Density (g/cm^3)	Diameter (mm)	Tensile strength (MPa)	Young's modulus (GPa)	Elongation at brake (%)
Flax	1.5	40–600	345–1500	27–39	2.7–3.2
Hemp	1.47	25–250	550–900	38–70	1.6–4
Jute	1.3–1.49	25–250	393–800	13–26.5	1.16–1.5
Kenaf	1.5–1.6	2.6–4	350–930	40–53	1.6
Ramie	1.5–1.6	0.0049	400–938	61.4–128	1.2–3.8
Curaua	1.4	7–10	500–1100	11.8–30	3.7–4.3
Abaca	1.5	10–30	430–810	31.1–33.6	2.9
Sisal	1.45	50–200	468–700	9.4–22	3–7
E glass	2.55	15-25	2000–3500	70–73	2.5–3.7

automotive industry, railways, aerospace and marine applications. Synthetic fibres absorb less moisture, corrosion resistant and have low manufacturing cost. Synthetic fibres are sourced from petroleum sources because of which they are not environmentally friendly, and hence, many researchers have explored natural fibres like hemp, jute, sisal in composites which not only provide high strength and stiffness but these are also biodegradable (Koronis et al. 2013). The cell structure of natural fibres itself is a composite structure where strong cellulose is reinforced on soft lignin and hemicellulose matrix which provides them high tensile strength as shown in Table 2 (Dicker et al. 2014). Although natural fibres are biodegradable and have high tensile strength, there are few negatives of natural fibres too. The natural fibres show variation in physical properties based on the season of harvesting, locality of the crop, part of crop and way of processing of biomass. Also, natural fibres have compatibility issues with many synthetic matrices and uniform dispersion is not obtained.

2.2.2 Particulate-Reinforced Composites

Particulates are particles reinforced on a matrix material. Particulates are either large particle composites which are formed by bonding large particles with soft matrix restricting the movement of the matrix or dispersion-strengthened composites with small particles with the size in the range of 10–100 nm. In dispersion-strengthened composites, matrix bears most of the load and small particles limit plastic deformation. Most common large particle composite is concrete which is made of cement matrix with bonds with particles of different sizes like gravel and sand. Cermets are also a very good example of large particle particulate composite wherein composite particles which are strong and brittle are bonded with the metal matrix which is soft and ductile (Woodford 2017). Cermets are used in cutting tools for hardening steels. Dispersion-strengthened particulate composites like reinforced rubber which contain carbon nanotubes reinforced on natural rubber are used in rubber tires and perform better than conventional rubber.

2.2.3 Structural Composites

Structural composites are special form of composites in which different materials with varied properties are brought together to form a material which compensates the weaknesses of the individual material. Most abundantly available laminate material present in nature is wood in which a softer lignin is embedded with the comparatively stronger cellulose which together forms the stiffness and the strength of the wood (Hinestroza and Netravali 2014). The structural composites are made of three main parts: two faces which are mostly stiff and thin with high tensile strength, middle core which is thick, lightweight and weak. All these materials are bonded together with the help of adhesives. Laminated composites are used in mainly structural applications. The structure in which the two stiff faces sandwich a weaker thick phase is called sandwich composite. The significance of the sandwich composites is that faces act as resistance to external bending movement like a stress couple, and the inner core resists the shear forces and stabilizes the faces against wrinkling or bulking (Ratwani 2010). The component of the sandwich laminates is the adhesive which bonds the core and faces together. If the bonds between the face and core are not strong, they might not be able to face the outer tensile stress. Structural laminates are used in cars, terrain vehicles, trains, etc., for the purpose of thermal insulation, reducing structural weight with keeping the strength intact, reduction in emissions and manufacturing costs. Along with application in automobiles, they are also included in houses as floors, panels, etc. Laminate composites are also type of structural composites which are formed by embedding multiple layers of sheets one above the another to form materials with enhanced tensile strength, low bending stiffness, improved thermal resistance, etc. (Mallick 1997). Laminated composites contain two or more layers, and their alignment is important in increasing their tensile strength and overall properties. Alignment of the layer in similar direction improves the overall stiffness of the composites. Plywood is one of the examples of such composites in which wood panels are bonded with fibre-reinforced plastics for stiff-laminated composites which have applications in house interiors. Laminated structural composites have application in fall ceilings, furniture making, ski complex, etc.

2.3 Advantages and Limitations of Composites

Advantages of Composites

1. High resistance to corrosion degradation and fatigue.
2. High strength-to-weight ratio. Composites have 25–45% lesser weight than conventionally used materials.

3. Composites need lesser structural repairs which significantly reduce maintenance cost.
4. Fibre reinforcements can be laid efficiently to sustain loads; i.e. they can be tailored to meet the requirements.
5. Composite panels show enhanced resistance to mechanical damages like dents.
6. In case of composites, it is easier to develop aerodynamic structures with smooth surface finishes.
7. Composites provide increased torsional stiffness because of which composites are useful in making structural and high-speed components.
8. Damage by impacts is highly resisted by composites compared to other materials.
9. Composites have low coefficient of thermal expansion along with low thermal conductivity which makes them dimensionally stable. Composite materials like carbon–carbon composites can be tailored to withstand high-temperature environments.
10. Composites have high corrosion resistance, durability and the weatherability because of which they reduce maintenance cost.
11. Metals are machined to required shape and configuration which leads to material wastage unlike composites wherein they are directly built to shape.
12. Composites provide improved wear properties and friction because of which some composites like carbon–carbon composites are used in airplane brakes due to their combined quality of heat stability and friction.

Limitations of Composites

1. Composites are costly mainly due to the high cost of fabrication and raw materials.
2. Some composites have weaker transverse properties.
3. Composites have weaker and softer matrices whose role is to support reinforcing material because of which they have low toughness.
4. Reusability or disposal of some composites is difficult as few are partially degradable, and incineration is the only other method used for their disposal which has environmental effects.
5. Composites are difficult to attach compared to metals.
6. Composites contain two different materials bonded together which makes their repair and analysis difficult.

Applications of Composites

1. Composites have applications in making aerospace structures. These are used in making military aircrafts, commercial airlines and space shuttle satellite systems.
2. Composites are used in marine machinery like boat bodies, kayaks.
3. Composites are used in making military armours, bulletproof vests, etc.

4. Composites also have biomedical applications and are used in making implants, biomedical devices, etc.
5. Automotive industry widely uses composites in making bumpers, car bodies and drive shaft.
6. Composites are used in tanks used for storing chemicals, piping, valves, etc., in chemical industries.
7. They are used in electrical devices as insulators, connectors, etc.

3 Effect of Composites and Their Materials on the Environment

Environmental impact of composites can be assessed by post-disposal impact of composite on environment and impact of manufacturing of matrix and reinforcing materials on the environment. Manufacturing of matrix materials like thermosets involves indoor air pollution by pollutants like VOCs, and similarly, disposal of non-biodegradable composites is hazardous to aquatic-like and other ecosystems.

3.1 Impact of Different Composite Materials on the Environment

Matrix materials are the important component of the composite which provides support and shape to the composite structure.

3.1.1 Impact of Polymers

Thermosetting polymers like epoxies, cyanate esters, phenolics, polyimides are widely used as matrix materials in composites. Thermosets are non-biodegradable polymers and have the adverse impact on aquatic environments. Toxic leachates in thermosets lead to water pollution and irreparable effect on the ecosystem. Manufacturing of thermosets also leads to the emission of volatile organic compounds which have hazardous health impacts and leads to air pollution. Thermoset polyesters involve emissions of air pollutants like particulates, nitrogen oxides, hydrocarbons, sulphur oxides, carbon monoxide and carbon dioxide; most of these pollutants are emitted during energy required for laundering process (Smith and Barker 1995). Unlike thermosetting polymers, thermoplastics can be recycled and remoulded which makes them a comparatively better option but thermoplastics have their own environmental impacts as they are not biodegradable and can stay in the ecosystem for millions of years. Petroleum-derived polymers also are obtained from non-renewable resources. Some thermoset resins can have toxic constituents

Table 3 Effect of polyvinyl chloride on environment based on life-cycle analysis (Ye et al. 2017)

Categories	Unit	Value
Climate change	kg CO_2	2.82×10^3
Ozone depletion	kg CFC-11	9.30×10^{-5}
Terrestrial acidification	kg SO_2	9.59
Freshwater eutrophication	kg P eq	0.03
Marine eutrophication	kg N eq	0.48
Human toxicity	kg 1,4-DB eq	428.42
Photochemical oxidant formation	kg PM_{10} eq kg NMVOC	12.29
Particulate matter formation	kg PM_{10} eq	3.58
Terrestrial ecotoxicity	kg 1,4-DB eq	14.99
Freshwater ecotoxicity	kg 1,4-DB eq	0.76
Marine ecotoxicity	kg 1,4-DB eq	6.29
Ionizing radiation	kBq U235 eq	110.41
Agricultural land occupation	m^2a	10.34
Urban land occupation	m^2a	7.20
Natural land transformation	m^2	0.17
Water depletion	m^3	57.52
Metal depletion	kg Fe eq	48.48
Fossil depletion	kg oil eq	1.12×10^3

which can lead to hazardous health and environmental impacts as shown in Table 3. Chemical processing steps to form composites lead to wastage of resin materials which if containing toxic chemicals lead to health hazards (Sands et al. 2001).

Thermoplastics like polyvinyl chloride (PVC) are used in composites as matrix materials. PVC is recyclable but a developing country like China hardly recycles 25% of the produced plastic, and this number can go lower in other developing countries (Ye et al. 2017). Elements like mercury and chlorine are mainly responsible for human toxicity and ecological toxicity. Production of PVC involves SO_x and NO_x emission of around 0.43 kg along with 0.21 kg of particulates. Impact of polyvinyl chloride on different categories is given in Table 3.

3.1.2 Impact of Metals

World produces one lakh kilo tons of steel and 30,000 kilo tons of aluminium which makes them the most abundant metal in the world (Norgate et al. 2007). Aluminium is used in making composites having applications in aerospace industry. Production of metals involves harmful emissions and also affects land and environment. Metal extraction sites are away from urban areas and mostly lead to noise pollution in remote areas affecting surrounding habitats. Metals like cadmium, lead and mercury are extremely toxic and can pollute land and water bodies (Norgate et al. 2007). Even metals like zinc and copper being biologically essential are highly toxic in large concentrations. Steel production processes lead to high amount of carbon dioxide emissions from processes like iron ore reduction, blast furnaces. Energy consumption in steel production industry is through combustion of coal which contributes to 90% of carbon dioxide emissions (Environmental Impact of the Processes 2017). NO_x, SO_x, dioxins and particulate matter are other pollutants emitted from steel industry in coking plants, electric arc furnaces, reheating and heat treatment furnaces.

3.2 Impact of Composites

Composites have mainly two components: reinforcing material and matrix material. Composites are the combination of two materials which give them unique properties. Problems with composites are that while considering degradation or disposal of composites, properties of both the composite components must be considered. In will to make greener composites, reinforcing materials like natural fibres reinforced on PVC or polyurethane are administered which lead to partially degradable materials which have disposal issues too. Composites can be divided into three types based on the environmental point of view and their biodegradability.

3.2.1 Non-Degradable Composites

Non-biodegradable composites have both the components non-biodegradable. Reinforcement material is mostly a synthetic fibre reinforced on a PVC matrix. Non-degradable composites if developed from thermoplastics can be recyclable and can reduce environmental impact. Major concern about non-degradable composites is their petroleum source. One of the examples of non-degradable composites is Kevlar-reinforced PVC.

3.2.2 Partially Degradable Composites

Partially degradable composites contain one of the components; mainly, matrix material is biodegradable. Partially degradable composites also contain reinforcing materials like jute, hemp, sisal fibres reinforced on a non-degradable matrix like polyurethanes. A major concern of partially degradable composites is that these are not recyclable and nor are they completely degradable because of which their disposal can only be done by gasification and incineration (Koronis et al. 2013). Partially degradable composites are used in automobile industry, furniture industry, gardening items, packaging, etc.

3.2.3 Biodegradable Composites

Biodegradable materials constitute biodegradable composites. Biodegradable composites contain both reinforcing material and matrix materials as environmentally degradable at normal environmental conditions. Biopolymers can be classified as agro-polymers obtained from agricultural sources, chemically synthesized materials from agro-based chemicals and chemicals made by microbes like polyhydroxybutyrates. Biopolymers like celluloses, starch, hemicelluloses are obtained from agricultural sources, and nanocelluloses have unique properties among these. Biodegradable polymer composites are still under research as biodegradable polymer composites are yet to show high strength-to-weight ratio and high-temperature resistance compared to conventional composite materials. Promising research has shown high scope for the development of high strength, high performance completely biodegradable composites in future (Table 4).

4   Environmentally Harmless Green Composites

Composites being a wonder material with superior properties have found wide applications in different industries like the automobile industry, aerospace industry, shipbuilding sector. In past decade or so, there is increasing concern about environmental impact of these composites having two or more different constituents. Green composites have been developed to produce environment-friendly composites. These are either completely biodegradable, or they are partially biodegradable. Recent developments in making completely degradable green composites like starch and bamboo fibre-based composites, soy protein-based composites with natural fibres show comparable properties (Lodha and Netravali 2002; Netravali and Chabba 2003).

4.1   Green Composite Reinforcements and Matrix

4.1.1   Natural Fibres

Natural fibres can be sourced from both plant and animal sources as shown in Fig. 1. Vegetable fibres contain polysaccharides, and animal fibres contain proteins; to use these fibres like reinforcements, certain constituents like hemicelluloses, lignin, wax, proteins need to be removed to increase fibre adhesion. The wood itself is a form of the laminated natural composite which makes it extremely strong; wood pulp is also used in making composite reinforcements (Hinestroza and Netravali 2014). Physical properties of these fibres are compared in Table 2 which shows natural fibres have the lesser density than glass fibres because of which they are considered the suitable alternative in automotive panel composites. But mechanical

Table 4 Comparison of biopolymers and synthetic polymers (Dicker et al. 2014)

Polymer type		Density (kg/m³)	Glass transition temperature (°C)	Melt temperature (°C)	Young's modulus (GPa)	Tensile strength (MPa)	Elongation (%)
Biopolymers	PLA	1210–1250	45–58	150–162	0.35–3.83	21–60	2.5–6
	PHB	1180–1260	4–15	168–182	3.5–4	24–40	5–8
	Starch	1000–1390	–	110–115	0.125–0.85	5–6	31–44
Synthetic	Epoxies	1110–1400	669–167	–	2.35–3.08	45–89.5	2–10
	Polyesters	1040–1400	147–207	–	2.07–4.41	41–89	2–2.6
	Polypropylene	890–910	0.9–1.55	–	0.9–1.55	28–41	100–600

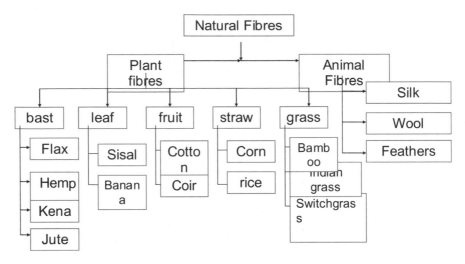

Fig. 1 Classification of natural fibres (Mohanty et al. 2002)

properties of natural fibres are comparatively lesser than synthetic fibres like Kevlar, aramid, and this shortcoming can be solved by increasing the density of the natural fibres (Mohanty et al. 2004). Natural fibres have some positives over synthetic fibres in terms of renewability, biodegradability, non-toxicity, good insulation properties and low machine wear. Natural fibres can substitute synthetic fibres in future if their water absorption property is reduced and fibre length is increased to increase the mechanical properties of the natural fibres.

Plant Fibres

Plant fibres are concisely discussed in the literature and are obtained from different parts of plants like bast, straw, leaf, fruit and grasses (Satyanarayana et al. 2009; Hills 2011; La Mantia and Morreale 2011). However, natural fibres have certain shortcomings compared to synthetic fibres and show variation in the product due to factors like age of the plant, geographical and climatic growth conditions, harvesting methods, purification technology.

Similarly, there is a significant difference in mechanical properties of natural fibres and synthetic fibres mostly due to a lesser length of the fibre in case of natural fibres. Water absorption property of natural fibres also reduced their adhesion because of which composite processing is a tedious job. Hydrophilicity of these natural fibres leads to the formation of clumps during processing.

Bast fibres are thoroughly exploited for making green composites, but now other vegetation is also considered for its study on green composites. Perennial grasses are single seed crops which need fewer amounts of fertilizers and water; hence, they are studied for composite materials. Switchgrass is studied for its potential as a reinforcing material for composites which showed that switchgrass PP composite showed better mechanical properties than jute fibre PP composite which shows considering different types of natural fibres can be essentially fruitful in developing green composites with enhanced properties (Zou et al. 2010).

Merits of lignocellulosic fibre-reinforced composites are (Satyanarayana et al. 2009)

1. In case of impact brittle fractures are formed, hence brittleness is reduced.
2. Enhanced performance at lower weight.
3. Increase in strength (25–30%) with no significant increase in weight.
4. Comparatively cheaper cost is lesser than the base resin.
5. Full recyclability is possible.
6. Up to 30% reduction in mould cycle time.
7. Processing of lignocellulosic composites is non-abrasive to the machinery.
8. Appearance of lignocellulosic composites is natural.
9. Lignocellulosic composites have low thermal expansion coefficient.
10. Good sound abatement capability is shown by lignocellulosic composites.
11. Reduced energy loss is observed in lignocellulosic composites as they have better energy management.
12. They have more shatter resistance.
13. Lower mould shrinkage is evident in lignocellulosic composite processing.
14. It is easy to colour them, hence low energy consumption in dying.
15. Lignocellulosic composites have high flex modulus which is up to five times the base resin.
16. They exhibit high tensile modulus around five times the base resin.
17. High notched impact which is up to two times the base resins is shown by lignocellulosic green composites.
18. Lignocellulosic composites need lower energy in processing.
19. Lignocellulosic composites easily meet the minimum recycling requirement and hence can be efficiently recycled.

Animal Fibres

Animal-based fibres have recently gained attention for reinforcing material for green composites; especially, traditional animal fibres like silk and wool have been tested (Kim et al. 2013; Surip et al. 2016).

Animal fibres have obtained interest due to being high-potential, high-performance fibres because of their surface toughness, flexibility, high aspect ratio and are less hydrophilic than other natural fibres. Animal fibres have different unique properties like keratin fibres possess significant volume of air resulting in low-density fibres with low dielectric constant which makes them suitable reinforcing material in electronic applications. Low fibre length is one of the reasons behind comparatively low tensile strength of the plant fibres, but some animal fibres like silk have long fibre length (1500 m) which gives them superior mechanical properties (Shao and Vollrath 2002). Mechanical strength of silk also strongly depends on the species of silk worn and method of spinning. Silk has low temperature resistance but has high antibacterial resistance and UV resistance. Silk composites have applications in biomedical applications in making grafts and tissue engineering application (Cheung et al. 2009). Animal fibres do not need fertile land and fresh water for growth which makes them a more sustainable alternative, but animal fibres can incur high cost and uncertain supply which is a shortcoming.

Fibres from Waste

Fibres from waste are more sustainable source of reinforcing material as it is part of an agricultural or industrial waste which is utilized for making a new material; this not only gives raw material at low cost but also provides waste disposal solution to the agricultural waste. Different agricultural and industrial wastes like sunflower stalk, rice husk, corn husk, wheat straw, soy stalk, sunflower stalk can be used as cellulosic fibre source for reinforcements (Zini and Scandola 2011). Wastes like corn husk are widely available from corn processing industries which also solve the problem of sourcing and geographical variation found in natural fibres. Rice husk and lingocellulosic biomass waste reinforced on polyethylene matrix have been studied to check their mechanical properties.

Animal processing industries like duck, ham and chicken fillet processing companies produce a huge amount of protein waste containing feathers, wool, etc; this waste can be used as reinforcements in composites. Chicken feather fibres are available in large amounts worldwide, chicken being one of the widely consumed meats globally. The hollow nature of these fibres leads to low-density fibres which can be used in automobile applications. Keratin fibres reinforced on PP are investigated and have showed increased stiffness in the polymer composite with low thermal resistance up to 200 °C (Barone et al. 2005). Waste utilization to obtain cellulosic- and keratin-based fibres requires initial processing of waste but also helps waste disposal along with providing sustainable reinforcement for green composites.

4.1.2 Biodegradable Polymers

Biodegradable polymers can be classified as agro-polymers, those obtained from agricultural sources like cellulose, microbial polymers obtained by microbial fermentation like polyhydroxyalkanoates, chemically synthesized by agro-based resource monomers and chemically synthesized by conventionally synthesized monomers. Biopolymer processing into composites involves lots of important factors like degradation time and temperature of the polymer along with their density which decides the weight of the final composite material (Baillie 2004). The stiffness and tensile strength of the biopolymers can be increased by reinforcing with the high tensile strength lignocellulosic fibres.

Classification of biopolymers (Satyanarayana et al. 2009)

1. Renewable sources

   I. Polysaccharides

      Cellulose, gums, chitosan, etc.

   II. Proteins

      Casein, collagen, gelatin.

   III. Lipids

      Cross-linked glycerides.

2. Chemical synthesized

   I. Polyacids.
   II. Polyvinyl alcohols.
   III. Polyesters.

3. Microbial synthesized

   I. Poly(hydroxyalkanoates).
   II. Bacterial cellulose.
   III. Xanthan, pullulan, curdlan.

4. Biopolymer blends

   I. Starch blends.
   II. Polyester blends.
   III. Caseinate blends.
   IV. Collagen/PVA blends.

Polylactic acid and starch are commercially used polymers for making coffee cups, lawn grasses, pens and razors. Polyhydroxyalkanoates are highly studied biopolymer because of its unique properties of high crystallinity and large spherulitic structures. Processing of starch blends by extrusion, injection, and blow moulding is possible, and starch behaves similar to synthetic resins and can also be moulded into thin films. Most biopolymers are water absorbing in nature, and thus, efforts are being made to improve their water resistance. In past decade, lot of research has been concentrated on using different monomers which can be used to make biopolymers water resistant.

Polylactic Acids (PLA)

Polylactic acids are one of the most promising biopolymers because of their high crystallinity and low rate of degradation. Polylactic acids are obtained from renewable sources mainly corn as corn is the most suitable starting material for PLA synthesis because of high purity product obtained though corn. The polylactic acid synthesis process is expensive because of a number of steps involved which makes it difficult to substitute it with commercially viable products.

Synthesis

PLA can be produced by two routes: one is ring polymerization of lactides in presence of Lewis acids and by a polycondensation reaction. Low-cost continuous PLA synthesis method is made by companies like Cargill Dow (Gruber and O'Brien 2005). Long chains of lactides can be obtained by ring polymerization, and molecular weight of the polymer depends on the reaction conditions and catalyst concentration. In industrial process, starch is converted into dextrose which is then used as a substrate in bacterial fermentation to obtain lactic acid. Lactic acid is converted into lactate by condensation, and further polymerization gives long-chain PLA biopolymer.

Properties

1. Good strength.
2. Film transparency.
3. Biocompatibility.
4. Availability from renewable resources.
5. Property variation can be obtained by varying polymers morphology and molecular weight.

Applications

PLAs biodegradation depends on temperature and moisture content, and hence, low molecular weight PLAs absorb but retain their shape so high molecular weight PLAs are preferred in different applications.

PLAs have higher costs compared to synthetic polymers. These high costs are mainly due to a number of steps involved in the production process along with the need to highly pure lactic acid to produce lactate monomer. PLA water resistance and transparency have lead to the use of PLAs in food packaging to promote environment-friendly material usage. The high cost of PLAs has given them a place in the biomedical application; in place of use of metals, PLAs are used in transplants and other biomedical applications. Cargill Dow Polymers, Hygail and Neste are companies which produce commercial products of PLA as shown in Table 5.

Table 5 Matrix reinforcement and processing of PHA, starch and PLA composites

Matrix	Reinforcement	Processing	Characteristics	References
Polyhydroxybutyrate copolymerized with polyhydroxy valerate	Recycled cellulosic fibres (10%)	Impregnation in fluidization chamber and hot pressing	Tensile strength: 130 ± 12 MPa (vol.%: 10) to 280 ± 48 MPa (vol.%: 25) Young's modulus: 6 ± 0.5 GPa (vol.%: 9.9) to 11 ± 0.14 GPa (vol.%: 27)	Wollerdorfer and Bader (1998)
PHB copolymerized with hydroxyvalerate	Steam exploded hemp	Twin screw extruder	After reinforcement of hemp, the tensile strength doubled to 30 MPa; around 3 GPa of quadrupled E-modulus was observed	Corbière-Nicollier et al. (2001)
PLA	Flax	Twin extruder	Strength and stiffness increased along with no degradation of PLA observed in compounding process	Oksman et al. (2003)
PLA	Flax	Compression moulding	Tensile strength and modulus increased, and highest tensile strength was observed in polyethylene glycol-grafted flax	Zini et al. (2004)
Biodegradable polyester	25% hemp	Twin screw extruder and injection moulding	Damage in extrusion and weaker fibre adhesion was observed	Keller et al. (2000)
Potato starch	Cellulose microfibrils from potato	Teflon mould casting	Better thermal stability and decreased water absorption with increased fibre content	Dufresne and Vignon (1998)
Thermoplastic starch	Sisal fibres	Press moulding/injection moulding	Increase in fibre content with marginal increase in flexural modulus Increase in tensile strength and Young's modulus was observed	Alvarez and Vázquez (2004); Alvarez et al. (2004a, b)
Wheat starch		Extrusion followed by injection moulding	Increase in storage modulus with well-embedded fibres. Better mechanical behaviour and improved thermal stability	Le Digabel et al. (2004)

Polyhydroxyalkanoates (PHA)

PHA is natural polyester produced by microorganisms which use them as energy storage. Most famous PHA is polyhydroxybutyrate, also known as PHBs. PHA has been biopolymer of interest mainly because of their similarity of mechanical properties with the synthetic polymer and their sourcing from renewable sources along with their biodegradability as shown in Table 5.

Synthesis

PHA is synthesized by a fermentation process with the help of bacteria *Alcaligenes eutrophus* for making products like polyhydroxybutyrate covalerate. The nature and amount of PHA produced depend on types of microorganisms as many microorganisms can produce PHA. The quality of product also depends on the substrate concentration, type of substrate, fermentation conditions, enzymes and coenzymes involved in catalysis of the process. Gene manipulation is essential to obtain the commercially viable dry weight of product in the continuous industrial process. Different carbon sources used in PHA fermentation process are propionic acid and other sources like pentanoic acid, 4-hydroxybutyric acids, 1,4-butanediol. After the fermentation process is over, the fermentation broth is extracted to obtain PHA. High-quality and high-quantity PHA can be obtained by continuous fermentation process with advanced separation technologies. The productivity of the PHA can be increased by using different carbon sources; especially, sugar hydrolysates from agricultural waste products which will also make this PHA production process sustainable (Yu et al. 1999; Yu 2001).

Properties

PHB is brittle in nature but can be made flexible and transparent by adding covalerates. Some researchers have claimed that the properties of PHB are similar to those of the polypropylene because of morphological similarities with polypropylene. PHB is susceptible to thermal degradation not much above its melting point which makes its processing difficult. PHB has the tensile strength of around 45 Mpa. PHB is biodegradable similar to other biopolymers, and its biodegradability depends on its surface properties and environmental conditions like temperature and moisture. Considerable effort has been put to improve the properties of PHB with the help of co-polymerization of hydroxyvalerates which improve the elasticity and transparency and reduce the brittleness of the polymer.

Applications

PHA has a low temperature of degradation as discussed earlier because of which the processing of the PHA is difficult as during processing high temperature will lead to degradation of PHA. The commercially available product which is made from PHA technology is BIOPOL as shown in Table 5. Biopolymers in PHA category were used in making cosmetic bottles and shampoo bottles. Looking at the properties and present applications, PHA can be used in making diapers, plastic bags and fast food containers (Baillie 2004).

Starch

Starch is an energy storage oligosaccharide present in different plants and microorganisms. Starch is made up of glucosidic units bonded together with glycosidic linkages. Starch contains around 80% amylopectin and 20% amylose (Baillie 2014). The proportion of these compounds varies based on environment and conditions. Starch occurs in the form of crystalline granules, and its crystallinity is due to the presence of amylopectin (Satyanarayana et al. 2009). Starch has poor high temperature stability, high water absorption and brittleness because of which plasticizer is essential for starch-based polymers to have commercial applications.

Properties

1. Short-term durability.
2. Destructurization can be done to obtain a thermoplastic material due to the presence of crystalline regions.
3. Starches with the high degree of destructuralization have elevated tensile strength and elongation.
4. Plasticizers like polyvinyl alcohol, citric acid, triethylene glycol, water urea can be used to enhance the properties of starch.
5. Water-soluble polysaccharides and proteins can be used as fillers to improve its mechanical properties.
6. The ratio of amylopectin and amylose, temperature and pressure in the extruder, moisture, etc., can be varied to change viscosity, water absorption of the starch-based thermoplastics.
7. Biodegradation of starch is common in many bacterial species, and amorphous regions offer more water accessibility and bacterial degradation compared to crystalline regions because of which separating crystalline regions can increase its shelf life.
8. Thermoplastic starch (TPS) shows degradation at high temperature during processing, and hence, using lubricants like silicone, lipids can be used to reduce processing temperature and increase the flow properties.

Applications

Starch is hydrophilic in nature because of which it is not used in food application, and acetylation of starch can increase water resistance to some extent. One more efficient way to increase water resistance is by making starch blends in other polymers like polycaprolactone. TPS has applications in packaging liners, agricultural mulch, loose fills, etc. Starch foams obtained from extrusion are used in trays and in a loose filling. Commercial starch is produced by Italy-based company named Novamont which sells starch granules (Bioplast) for moulding, starch films (Bioflex) and foams (Biopur) (Table 6). Other manufacturers who make starch-based products are Midwest Grain Products, Earth Shell, Hayashibara Chemical Labs, AVEBE, National Starch, Rodenburg Biopolymers, etc. List of commercially available biopolymer given in Table 6 is not exhaustive, and there are many more such companies who produce biopolymers.

Table 6 Commercially available biopolymers and their companies (Mohanty et al. 2000)

Material class	Manufacturing	Product name
Cellulose acetate	Mazzucchelli, Planet Polymer	Bioceta, EnviroPlastic
Copolyester	BASF, Eastman	Ecoflex, Easter bio
Polycaprolactone (PCL)	Birmingham Polymers, Planet Polymer, Solvay	Poly(e-caprolactone), EnviroPlastic, CAPA
Poly(ester amide)	Bayer	BAK 1095, BAK 2195
Poly(ethylene terephthalate)-modified	DuPont	Biomax
Polyglycolide (PGA)	PURAC, Alkermes	PURASORB
Polyhydroxyalkanoates (PHAs)	Metabolix, Biomer	PHA, Biomer
Poly(lactic acid) (PLA)	Cargill Dow Polymers, Hygail, Neste	ECO PLA, PLA, poly (L-lactide)
Poly(vinyl alcohol) (PVOH)	Novon, Planet Polymer	Poly-NOVON, Aquadro
Starch and starch blends	AVEBE, BIOTEC, Novon, Novamont	Paragon, Bioplast, Poly-NOVON

Cellulose

Cellulose is abundantly found in almost all plant materials, especially in wood. Most of the carbon present on earth is in the form of cellulose (Hinestroza and Netravali 2014). Cellulose can be either extracted from plant sources or they can also be produced in a bacterial fermentation process. Cellulose extracted from plant sources has amorphous and crystalline regions. The amorphous region in cellulose is removed to enhance its mechanical and thermal properties, and obtained cellulose is called microcrystalline cellulose. Reduction in particle size of cellulose to nanosize leads to the formation of cellulose with enhanced tensile strength and Young's modulus, and this form of cellulose is called nanocrystalline cellulose. Cellulose is also produced by bacteria of acetobacter spp. And *Acetobacter xylinum* is the most studied organism for the production of bacterial cellulose. Cellulose has tensile strength and Young's modulus comparable with any of the synthetic fibres and if combined with suitable bio-based resin cellulose can be a renewable substitute to any petroleum-based synthetic composites.

Extraction of Cellulose

Cellulose microfibrils are located inside plant sources in combination with hemicelluloses and lignin. Cellulose fibrils are entangled between matrix made of soft lignin and hemicelluloses in naturally occurring laminates like wood. It is essential to select proper treatment to extract cellulose microfibrils so that their structure is retained and the highly crystalline product is obtained. Initial fibrillation studies involved mechanical treatments mainly for the extraction of celluloses. Cellulose extraction method can be either a top-down approach wherein mechanical and shear forces are applied to obtain microfibrillated cellulose. Different treatments like high-pressure homogenization, super-grinding method and cryocrushing reduce cellulose fibrils in longitudinal form and avoid transversal cutting to obtain optimum fibril structure. Combination of the above-mentioned techniques is used to obtain cellulose of desired morphology. Microfluidization is a process in which a fluid stream containing cellulose is suspended to extreme shear to reduce cellulose fibril length. Processes like electrospinning involve bottom-up approach wherein charger polymers are spun into fibrils in range of few hundred nanometres by applying electric charge (Szczesna-Antczak et al. 2012).

Cellulose is also produced by the bacterial fermentation process in which acetobacter species bacteria are cultivated in fructose- or glucose-based fermentation media, and after the span of 7 days, a cellulose film starts appearing in the batch culture which can be harvested and sterilized to obtain pure cellulose. Bacterial cellulose film mainly contains water, and hence, low productivity is one of the shortcomings of bacterial cellulose method along with reactor designing constraints.

Applications of Cellulose Nanofibrils

Nanofibrils have high scope to have applications in different sectors like medical, cosmetics, environment, energy, electronics textiles. They can be in drug delivery as drug carriers, prosthetics, dressings, surgical materials, etc., in the medical sector. High surface area and absorptivity of cellulose fibrils make them promising candidate in the cosmetic industry to be used in creams, masks, etc. Cellulose fibrils can also be used in electrical equipment like computers, shields for electromagnetic radiation. Cellulose fibrils have applications in textiles clothing, etc. Adsorbers, nanofilters, sensors, filters, etc., can also be made from cellulose fibrils (Szczesna-Antczak et al. 2012). Most of the above applications are mentioned in the research stage, and very few products made from nano-fibrillated cellulose are available in the market; hence, cellulose might be a sustainable substitute to petroleum-based polymers only if high productivity and low-cost processes are developed for cellulose nanofibril production.

4.2 Processing Aspects of Green Composites

Injection moulding, compression and extrusion are the most common processing techniques which are used in the processing of the composites. Green composites involve materials which have low melting points, and hence, degradation during processing is possible. Similarly, the presence of hydrophilic components makes processing process difficult. Although there are few challenges to overcome to develop efficient processing techniques to suitably processing green composites, there are few merits which processing of biodegradable materials offers like less abrasion of processing components and tools, the absence of airborne particles which are common in glass fibre reinforcements because of which many respiratory problems arise.

Factors essential for processing of green composites include

1. Proper mixing of natural fibres.
2. Selection of appropriate biopolymer matrix.
3. Essential surface treatment is needed mainly if the nature of matrix and rein-forcements are different.
4. Low-cost processing techniques which do processing at higher speed without affecting the product quality.

Different processing techniques specific to the nature of biodegradable polymers and fibres should be developed to solve issues like mixing degradation and attrition which lead to reduction in the mechanical properties and fibre distribution. Development of techniques will lead to good wettability, reduced fibre degradation during processing; high volume fraction of natural fibre incorporation with opti-mum fibre orientation can be obtained. Techniques like dry powder impregnation can be used which negate the use of solvents in the process which have their own environmental implications, ease mixing process, provide uniform dispersion of components having more than one phase, consume lower amount of energy, and hence, energy efficiency is attained and recycling of powders to be impregnated can be possible. Powder impregnation process also comes with some disadvantages like the reinforcing fibres have to be in the powder form which can reduce the strength which is the major purpose of reinforcing material. Different processes are used in making green composites which mainly include filament winding, layup methods, resin transfer moulding, injection moulding and autoclave bonding (Kandpal et al. 2015).

4.2.1 Filament Winding

Continuous fibres impregnated with resins are wound around a rotating mandrel that has shape for the desired product, and the obtained resin is cured and treated to obtain product of desired consistency (Ratwani 2010). Immediate removing of fibre roving from resin bath avoids forming of helical pattern on the mandrel. This process of filament winding is repeated several times, and each winding has criss-cross pattern with the previous layer. Filament winding can be repeated to adjust the thickness and consistency of the end product. Filament winding process can be either continuous winding process or discontinuous winding process depending on the type and nature of desired product. Continuous winding process is used to manufacture large-to-small piping parts mainly used to transfer media like water, solvents. Discontinuous winding process produces product at lesser rate compared to continuous winding process. Discontinuous winding process is used to manufacture high-pressure vessels, high-pressure reactor parts, so parts with high thickness and strength are developed by continuous winding process.

4.2.2 Contact Moulding

Contact moulding is also called as hand layup method. Contact moulding is one of the oldest methods of moulding which has comparatively simpler process than other types of moulding, especially for making fibreglass-reinforced composites. Contact moulding is used to make different products like wind turbine blades, boats. In contact moulding process, the mould is treated with mould release agent and further thin gel (resins) coat is applied on the mould and allowed to set. Once the gel is set, the resins and fibres are applied in layers; fibres are mostly in the form of mats or films. The layers are rolled to remove any air between them. After the layers are set, the part is cured and then the fully hardened composite is removed carefully (Kandpal et al. 2015). One of the positives of the contact moulding is the fine aspect of layer obtained by this method, lower cost of moulding and easy implementation of the process (Satyanarayana et al. 2009). Few negatives of this process are only one smooth face is obtained, quality of the product is dependent on the skill of the person doing moulding, and a large amount of products are difficult to produce by this method.

4.2.3 Resin Transfer Moulding (RTM)

Resin transfer moulding is the process of forming composites in which two moulds are used from two sides because of which two sides of panels are formed. In resin transfer moulding, both the surfaces can be obtained with smooth layer unlike contact moulding. The two sides fit together to form mould cavity in which the reinforcement material is already added. The material used to make this mould is mainly aluminium and steel, but sometimes composite materials are also used in resin transfer moulding (La Mantia et al. 2008). This process can be carried out in ambient and elevated temperatures. In this process, the reinforcing material is already present in the mould cavity and resin matrix is infused from outside. The method of infusion and condition involved is also different. The following are different positives of resin transfer moulding:

1. The surface quality of the obtained product is good.
2. Large and complex shapes can be made by this method.
3. This process provides dimensional tolerances.
4. RTM process needs low capital investment.
5. Lesser material wastage is observed at the end of RTM processed.
6. This process flexibility of tooling.
7. Less number of labours is needed in this process.
8. Inserts and reinforcements can be added at the point of infusion which provides greater strength.
9. Absolutely zero air is trapped in the product.
10. High production rate can be obtained by process automation.

4.2.4 Injection Moulding

Injection moulding process is used to make low-cost products in large numbers which are mainly plastics. Though thermosets can also be used in injection moulding, low-cost thermoplastic materials are most probably produced by this process. In injection moulding process, the resins and reinforcements are heated so that resins plasticize and form into a thick viscous solution which is then passed through an injector and injected inside a mould (Kandpal et al. 2015). Final finishing might not be required in this process as outer colourants, etc., and can be injected to obtain the finished product. The following are the benefits of injection moulding process:

1. Suitable process when intricate parts are to be produced in large quantities.
2. One moulded part produced by this process can replace assembly of components.
3. Parts can be moulded along with surface and colour finish; hence, finishing step is eliminated.

Injection moulded parts are thinly walled as mainly thermoplastics are used; hence, these products are not recommended for high strength applications. Thermoplastics are often reinforced with glass fibres to increase their tensile strength and can be used in heavy applications.

4.2.5 Autoclave Bonding

Autoclave bonding process contains two-sided mol set: one side is rigid mould, and another side is flexible mould made of silicone or extruded polymer like nylon (Ratwani 2010). Reinforcement materials like continuous fibre forms can be placed both manually and in continuous processes. In most cases, the reinforcements are impregnated in a resin film or resin film is kept on the moulds, and further reinforcements are kept, and upper mould is kept along with the application of vacuum so that the matrix and reinforcements are impregnated with no presence of air. Further, the moulds are kept in the autoclave which offers high pressure and temperature which leads to the formation of the end product. The advantage of this process is that high fibre loading can be attained in this process along with high structural efficiency. High strength materials like aircraft parts, marine and space crafts, and missiles etc are made by autoclave bonding process.

Advantages of Autoclave Bonding Process

1. High volume fraction of the reinforcements can be obtained in composite pressing method.
2. Thermoplastic and thermosetting polymers can be processed by this method.
3. A vacuum is formed during the process; no void spaces are formed in the end production.

4. Good control over resins and reinforcements is obtained along with high uniformity of the product and good adhesion between the layers.
5. Wetting of fibres is completely achieved.
6. A vacuum is used during processing better bonding of the core, and inserts are possible in autoclave bonding process.

4.3 Thermoplastic and Thermosetting Green Composites

Fibre reinforcements on matrix material can be enhanced by methods like chemical grafting wherein a polymer is attached to the fibre whose solubility matrix is similar to the matrix polymer. Polyethylene glycol is used as inter-facial polymeric material in many cases to reinforce lignocellulosic fibres on polymer matrix (Zini and Scandola 2011). Whenever a fibre is attached with inter-facial polymeric material to enhance the bonding between matrices and reinforcing material, it is essential that inter-facial material reacts only with the outer surface of the natural fibre material; otherwise, there will be loss of the mechanical properties of the reinforcing fibres which will do no good to the purpose of forming a composite.

Thermosetting Green Composites

Natural oils like soybean, castor and linseed oil can be polymerized to obtain resins which can be used as the substitute to petrochemical-derived polymers like epoxy resins, vinyl esters. Vegetable oils are renewable, naturally available and low cost which make them the suitable substitute for the conventionally available composites. Natural oils are triglycerides which can be polymerized when chemical functionalities are added to them. Different processes like open ring reaction with haloacids, ozonolysis, hydration and epoxidation are used in the production of thermosetting composites. Vegetable oil is obtained from plant sources, and hence, their production process has much lesser environmental impact than the petrochemical oil-based resins (Baillie 2004) (Table 7).

Table 7 Different thermosetting resins' green composites (Zini and Scandola 2011)

Composite manufacturing process	Fibre	Matrix	Reference
Resin transfer moulding (vacuum assisted)	Chicken feather	β-styrene + acrylated and epoxidized soybean oil	Hong and Wool (2005)
Resin transfer moulding (vacuum assisted)	Chicken feather	Pentaerythritol glyceride maleates of soybean oil	Hong and Wool (2005)
Resin transfer moulding	Hemp	Diphenylmethane diisocyanate + PCL + castor oil	Lee et al. (2009)
Sheet moulding compound	*Luffa cylindrica*	Diphenylmethane diisocyanate + castor oil	Gadhave et al. (2017)
Compression moulding	Flax	Methacrylated soybean oil + styrene	Melo et al. (2008)

Newly developed oil-based plastics are used as matrix in multiple composites and have applications in different sectors like making agricultural equipment, automotive components, making marine structures like pipeline and offshore equipments, civil infrastructure applications like bridges and highway components, railway applications like boxcars, carriages and hoppers, applications in construction industry like particle boards, engineering lumber and ceilings (Zini and Scandola 2011). Manufacturing of thermoset biocomposites includes different types of processes which are similar to those used in making conventional composites mould methods like resin transfer, compression moulding and vacuum infusion. The resin material is infused in reinforcing fibre mats, and further curing is done at elevated temperatures by applying vacuum or by using compression moulding. These composite manufacturing methods allow high fibre loading in resin moulds mainly because of low viscosity of the matrix material. Bio-based thermosetting composites are still scarce in the market mainly due to less availability of the thermoset resins in the market. Biocomposites made from epoxidized soybean oil having applications in non-structural indoor laminates are manufactured by Environ Biocomposites (USA) (Zini and Scandola 2011). Many such companies are developing thermoset composites which are available commercially.

Thermoplastic Green Composites

Polylactic acids and polyhydroxybutyrates are broadly used thermoplastic polymers for making biodegradable green composites. PLAs are obtained by the fermentation process and are used as matrix mostly along with nano-fibrillated carbons dispersed on PLA matrix. PLAs can be processed like plastics with processes like compounding and injection moulding. High tensile strength in composites can be obtained by increasing fibre load in composites. In PLAs, it is difficult to load large volumes of fibres in the matrix. This shortcoming is because large volume dispersion of reinforcing material is not uniform in the high viscosity of matrix materials. Similarly, it is difficult to reinforce long fibres in bio-based thermoplastic matrix by compounding and injection moulding; hence, short fibre reinforcement leads to lesser tensile strength. For longer fibre reinforcement, compression moulding needs to be used.

Commercially available thermoplastic composites like PLA-based composites are marketed by several companies globally mainly because of their low carbon footprint and environmental impact (Baillie 2004; Hinestroza and Netravali 2014). The wood-reinforced polylactic acid compound named FIBROLON is developed by a German company named FkuR which can be used in the panel, complex profiles, automobile parts, etc. Similar such applications and commercially sold composites are shown in Table 8.

Thermoplastic resins are used with different types of fibres like hemp, wood flour, flax and studied extensively by researchers; some of the recent studies are briefly shown in Table 9.

Table 8 Commercially available bio-based thermoplastics composites

Company	Product	Description
Kareline Oy (Finland)	Kareline PLMS	Biodegradable fibre-reinforced composites
GreenGran (Germany)		Wood-reinforced PLA
FASAL WOOD KG (Austria)	Fasal1	Granulates from PLA and wood waste for injection moulding
FuturaMat (France)	Biofibran	Wood and biodegradable polyesters

Table 9 Recently developed thermoplastic resins based green composites (Zini and Scandola 2011)

Composite manufacturing process	Fibre	Matrix	Reference
Compression moulding and injection moulding	Hemp	Cellulose acetate	Mohanty et al. (2004)
Injection moulding	Wood	Thermoplastic starch	Avérous and Boquillon (2004)
Melt mixing	Wood flour	Thermoplastic starch blend	La Mantia et al. (2008)
Injection moulding	Flax	PHBH	Zini et al. (2007)
Compression moulding and injection moulding	Flax	PHB, PHBV	Barkoula et al. (2010)
Compression moulding	Flax	PHB, thermoplastic starch, PLA	Bodros et al. (2007)

4.4 Attributes of Green Composites

4.4.1 Mechanical Properties

Natural fibre and glass fibre have comparable strength and stiffness, but when it comes to mechanical properties of composites big difference is observed. Major reason behind this difference is low volume fraction of unaligned reinforcements in green composites. One more reason behind low strength in composites is mainly due to hydrophilic natural fibres and hydrophobic synthetic polymers which are difficult to bind together and form strong composite material with high specific strength. Chemical treatment of fibres with chemicals like alkali can help reduce water absorption of the fibres which increases fibre–matrix adhesion as reported by Van de Weyenberg et al. (2006). Although efforts are being made to improve the overall strength of green composites, conventional composites still have better mechanical properties.

4.4.2 Varying Properties of Fibres

Natural fibres show variation in different properties like fibre length, chemical composition and fibre shape. These variations are due to variation in climate, harvesting timing, seed density, quality of soil, fertilization and location of fibre on the plant (Dittenber and Gangarao 2012). This variability is shown in Table 2 wherein wide variety in properties is evident in different types of fibres. Different efforts are being made to reduce the variability in fibre properties by understanding the effect of reducing variation in cultivation conditions. Upcoming technologies like hydroponics can be effective in growing fibres with similar properties. Even if we develop technology to grow fibres with similar properties in different places, short fibre length and fibre alignment are responsible for low tensile strength in natural fibres. Efficient approaches to develop long-chain fibres are essential to develop high tensile strength fibre.

4.4.3 Renewability

Plastics used in composites are sourced from petroleum sources which were formed thousands of years ago. These petroleum products emit lots of carbon dioxide in environment while production and during their disposal. Plastics from petroleum sources are unsustainable considering amount of time they take to decompose (Ye et al. 2017). Green composites mainly are sourced from plants, and hence, growing of green composite materials leads to reduction in carbon dioxide. Similarly, green composites are biodegradable; hence, green composites have no disposal problems compared to conventional composites.

4.4.4 Low Embodied Energy

Green composites need lesser energy for production compared to synthetic composites. Green composites contain ten times lesser energy embodied for production than synthetic fibres production (Dicker et al. 2014). Natural fibre production needs around 20–50% lesser energy compared to synthetic fibre production (Mohanty et al. 2001; Dicker et al. 2014). Hence, when it comes to energy embodied for production and carbon dioxide emissions green composites have lesser environmental impact.

4.4.5 Biodegradability

Materials degradation by living organisms is biodegradability, and material degradation can also be obtained by biological material like hydrolytic enzymes. Natural polymers have hydrolysable bonds in their backbone which makes their breakdown possible by living organisms. Some chemicals like polyesters and polyanhydrides are biodegradable as they have hydrolysable bonds. Synthetic plastics from petroleum sources are not biodegradable and can stay in the environment for years. Green composites being biodegradable have a big advantage over composites, and development of green composites with high mechanical properties will make them a greener substitute (Satyanarayana et al. 2009).

4.4.6 Low Cost

Cost difference between biopolymers and synthetic polymers depends highly on petroleum prices. Around a decade ago, cost of biopolymers is significantly higher than synthetic polymers. Since then due to the increase in petroleum prices, significant reduction in price gap between biopolymers and synthetic polymers is observed (Mohanty et al. 2000). Due to the economic change in past decade, there is 111% increase in polypropylene cost and 73% reduction in PLA cost; hence, biopolymers are expected to become cost-effective with time also due to the development of new technologies to make cost-effective biopolymers (Mohanty et al. 2000). Although the cost of biopolymers has reduced in the past decade, persistent efforts of high strength biocomposites need high fraction volume of fibres per composite which increases the cost of production of the biocomposites. Cost of fibre has reduced, but significant processing is essential to reduce water absorption properties of the biocomposites which adds to their cost. Hence, although there is reduction in the cost of biopolymers their processing incurs fair amount of cost.

4.4.7 High Water Absorption

Cellulose is an important constituent of biopolymers, and it is hydrophilic in nature which makes most of the biopolymers hold a high amount of water. High water-holding capacity is not a good asset for making composites as it leads to delamination, surface roughening and reduction in overall strength (Dicker et al. 2014). Durability and shelf life of the biopolymers also reduce due to the presence of water as there are chances of bacterial and fungal growth. Furthermore, synthetic polymers are mostly hydrophobic in nature on contrary to biopolymers because of which their compounding becomes more difficult. Presence of moisture significantly affects the mechanical properties also; the mechanical properties like tensile strength of composites made by biopolymers reduce when in contact with water.

4.4.8 Poor Durability

Green composites have poor durability mainly due to their high water absorption property. Green composites are more prone to fungal and bacterial growth and further environmental degradation of the composite. This leads to low shelf life of composites. Study on fungal growth on flax fibres showed growth in hardly 3 days which is a very short span and shows low durability of green composites (Stamboulis et al. 2000).

4.4.9 Non-Toxicity

Natural fibres are non-toxic in nature compared to synthetic fibres. This property of natural fibres makes them environmentally non-hazardous, and hence, environmental concerns and health hazards regarding the processing and production of biocomposites are lesser than synthetic polymers.

4.4.10 Biocompatibility and Bioactivity

Biocompatibility is the property of material to be in harmony with environment and biological organisms. Biocompatible materials do not create biological response like allergies. All the biopolymers like PHB, PLA, chitin have proven biocompatibility.

4.4.11 High-Temperature Degradation

Green composites have low processing temperatures and start degrading at 180–210 °C. High-temperature applications like engine parts or high-temperature reactor components cannot be made with green composites. High-temperature degradation reduces the applicability of green composites.

5 Conclusion and Future Prospects

Green composites have great future prospects considering their sustainability and environmental friendliness. But there are still few roadblocks in making green composites completely substitute conventional petroleum-based composites. Biopolymers have low thermal degradation temperature and due to their hydrophilic nature are difficult to bind with hydrophobic reinforcements which lead to difficulty in the processing of green composites. Partially degradable green composites leave non-degradable component behind, and their recycling is also difficult which makes their disposal difficult and few processes like incineration and gasification can be used to dispose them. Hence, using natural fibres reinforced on biodegradable polymer matrix like PLA are most environmentally sustainable alternative but these composites have lower tensile strength because of low fibre length, hydrophilicity, etc. Fibre length and morphology of fibres vary based on the location and climate where it is grown because of which there is variation in fibres available in different areas. New materials like cellulose nano-fibrillated green composites, carbon–carbon composites seem to have excellent properties and have high scope in future. So, green composites have great potential and new advances are coming up on regular basis. However, still a lot of research is required to produce high temperature and high tensile strength green composites to completely substitute conventional composites and to reduce environmental impact caused by these petroleum-based composites and their manufacturing processes. Future markets will definitely see green composites completely competing with synthetic composites. Nowadays, green composites have applications in packaging sector but future green composites will also have structural applications as well. Fibre reinforcements will be substituted by nanocrystalline cellulose crystals obtained from agricultural waste embedded in biodegradable matrix, and these future nanocomposites will not only have higher tensile strength but also they will exhibit sustainability and renewability. Carbon–carbon-based composites which have applications in aerospace industry and are costly for low-end products will be available cheaply because of low-cost synthesis and processing of composites. Hence, in future green composites will provide a cost-effective and comparable alternative to synthetic composites.

References

Alvarez VA, Vázquez A (2004) Thermal degradation of cellulose derivatives/starch blends and sisal fibre biocomposites. Polym Degrad Stab 84(1):13–21. https://doi.org/10.1016/j.polymdegradstab.2003.09.003

Alvarez VA et al (2004a) Melt rheological behavior of starch-based matrix composites reinforced with short sisal fibers. Polym Eng Sci 44(10):1907–1914. https://doi.org/10.1002/pen.20193

Alvarez VA, Kenny JM, Vázquez A (2004b) Creep behavior of biocomposites based on sisal fiber reinforced cellulose derivatives/starch blends. Polym Compos 25(3):280–288. https://doi.org/10.1002/pc.20022

Avérous L, Boquillon N (2004) Biocomposites based on plasticized starch: thermal and mechanical behaviour. Carbohyd Polym 56(2):111–122. https://doi.org/10.1016/j.carbpol.2003.11.015

Baillie C (ed) (2004) Green composites polymer composites

Baillie C (2014) Green composites polymer composites and the environment. Rev Adv Mater Sci 37(1–2):20–28. https://doi.org/10.2991/icmemtc-16.2016.148

Bajpai PK, Singh I, Madaan J (2014) Development and characterization of PLA-based green composites: a review. J Thermoplast Compos Mater 27(1):52–81. https://doi.org/10.1177/0892705712439571

Barkoula NM, Garkhail SK, Peijs T (2010) Biodegradable composites based on flax/polyhydroxybutyrate and its copolymer with hydroxyvalerate. Ind Crops Prod 31(1):34–42. https://doi.org/10.1016/j.indcrop.2009.08.005

Barone JR, Schmidt WF, Liebner CFE (2005) Compounding and molding of polyethylene composites reinforced with keratin feather fiber. Compos Sci Technol 65(3–4):683–692. https://doi.org/10.1016/j.compscitech.2004.09.030

Bodros E et al (2007) Could biopolymers reinforced by randomly scattered flax fibre be used in structural applications? Compos Sci Technol 67(3–4):462–470. https://doi.org/10.1016/j.compscitech.2006.08.024

Chawla KK (ed) (2013) Ceramic matrix composites. Springer Science & Business Media

Cheung HY et al (2009) Natural fibre-reinforced composites for bioengineering and environmental engineering applications. Compos B Eng 40(7):655–663. https://doi.org/10.1016/j.compositesb.2009.04.014 Elsevier Ltd.

Composite Matrix Materials (no date) Available at https://www.azom.com/article.aspx?ArticleID=9814. Accessed 16 May 2018

Corbière-Nicollier T et al (2001) Life cycle assessment of biofibres replacing glass fibres as reinforcement in plastics. Resour Conserv Recycl 33(4):267–287. https://doi.org/10.1016/S0921-3449(01)00089-1

Dicker MPM et al (2014) Green composites: a review of material attributes and complementary applications. Compos A Appl Sci Manuf 56:280–289. https://doi.org/10.1016/j.compositesa.2013.10.014

Dittenber DB, Gangarao HVS (2012) Critical review of recent publications on use of natural composites in infrastructure. Compos A Appl Sci Manuf 43(8):1419–1429. https://doi.org/10.1016/j.compositesa.2011.11.019

Dufresne A, Vignon MR (1998) Improvement of starch film performances using cellulose microfibrils. Macromolecules 31(8):2693–2696. https://doi.org/10.1021/ma971532b

Environmental Impact of the Processes (2017) Available at https://www.jernkontoret.se/en/the-steel-industry/production-utilisation-recycling/environmental-impact-of-the-processes/. Accessed 21 May 2018

Eriksen M et al (2014) Plastic pollution in the world's oceans: more than 5 trillion plastic pieces weighing over 250,000 tons afloat at sea. PLoS ONE 9(12):1–15. https://doi.org/10.1371/journal.pone.0111913

FAO (2009) How to feed the world in 2050, insights from an expert meeting at FAO. https://doi.org/10.1111/j.1728-4457.2009.00312.x

Gadhave RV, Mahanwar PA, Gadekar PT (2017) Bio-renewable sources for synthesis of eco-friendly polyurethane adhesives—review. Open J Polym Chem 07(04):57–75. https://doi.org/10.4236/ojpchem.2017.74005

Gruber P, O'Brien M (2005) Polylactides "NatureWorks® PLA". Biopolymers Online. https://doi.org/10.1002/3527600035.bpol4008

Hills P (2011) Recent advances in green composites. 425(2010):107–166

Hinestroza J, Netravali AN (2014) Cellulose based composites: new green nanomaterials. https://doi.org/10.1002/9783527649440

Hong CK, Wool RF (2005) Development of a bio-based composite material from soybean oil and keratin fibers. J Appl Polym Sci 95(6):1524–1538. https://doi.org/10.1002/app.21044

Jose JP et al (2012) Advances in polymer composites: macro- and microcomposites—state of the art, new challenges, and opportunities. Polym Compos 1:1–16. https://doi.org/10.1002/9783527645213

Kandpal BC, Chaurasia R, Khurana V (2015) Recent advances in green composites—a review. Glob J Pharmacol 9(3):267–271. https://doi.org/10.5829/idosi.gjp.2015.9.3.94289

Keller A et al (2000) Degradation kinetics of biodegradable fiber composites. J Polym Environ 8(2):91–96. https://doi.org/10.1023/A:1011574021257

Kim NK, Bhattacharyya D, Lin RJT (2013) Multi-functional properties of wool fibre composites. Adv Mater Res. https://doi.org/10.4028/www.scientific.net/AMR.747.8

Koronis G, Silva A, Fontul M (2013) Green composites: a review of adequate materials for automotive applications. Compos B Eng 44(1):120–127. https://doi.org/10.1016/j.compositesb.2012.07.004 Elsevier Ltd.

La Mantia FP, Morreale M (2011) Green composites: a brief review. Compos A Appl Sci Manuf 42(6):579–588. https://doi.org/10.1016/j.compositesa.2011.01.017 Elsevier Ltd.

La Mantia FP et al (2008) Effect of the processing on the properties of biopolymer based composites filled with wood flour. Int J Mater Form 1(SUPPL. 1):759–762. https://doi.org/10.1007/s12289-008-0286-7

Le Digabel F et al (2004) Properties of thermoplastic composites based on wheat-straw lignocellulosic fillers. J Appl Polym Sci 93(1):428–436. https://doi.org/10.1002/app.20426

Lee N et al (2009) Characterization of castor oil/polycaprolactone polyurethane biocomposites reinforced with hemp fibers. Fibers Polym 10(2):154–160. https://doi.org/10.1007/s12221-009-0154-1

Lodha P, Netravali AN (2002) Characterization of interfacial and mechanical properties of "green" composites with soy protein isolate and ramie fiber. J Mater Sci 37(17):3657–3665. https://doi.org/10.1023/A:1016557124372

Mallick PK (ed) (1997) Composites engineering handbook. CRC Press, New york

Melo BN et al (2008) Eco-composites of polyurethane and Luffa aegyptiaca modified by mercerisation and benzylation. Polym Polym Compos 16(4):249–256

Mohanty AK et al (2004) Effect of process engineering on the performance of natural fiber reinforced cellulose acetate biocomposites. Compos A Appl Sci Manuf 363–370. https://doi.org/10.1016/j.compositesa.2003.09.015

Mohanty AK, Misra M, Hinrichsen G (2000) Biofibres, biodegradable polymers and biocomposites: an overview. Macromol Mater Engi 276–277:1–24. https://doi.org/10.1002/(sici)1439-2054(20000301)276:1<1::aid-mame1>3.0.co;2-w

Mohanty AK, Misra M, Drzal LT (2001) Surface modifications of natural fibers and performance of the resulting biocomposites: an overview. Compos Interfaces 8(5):313–343. https://doi.org/10.1163/156855401753255422

Mohanty AK, Misra M, Drzal LT (2002) Sustainable bio-composites from renewable resources: opportunities and challenges in the green materials world. J Polym Environ 10(1–2):19–26. https://doi.org/10.1023/A:1021013921916

Netravali AN, Chabba S (2003) Composites get greener. Mater Today 6(4):22–29. https://doi.org/10.1016/S1369-7021(03)00427-9

Norgate TE, Jahanshahi S, Rankin WJ (2007) Assessing the environmental impact of metal production processes. J Clean Prod 15(8–9):838–848. https://doi.org/10.1016/j.jclepro.2006.06.018

Oksman K, Skrifvars M, Selin JF (2003) Natural fibres as reinforcement in polylactic acid (PLA) composites. Compos Sci Technol 63(9):1317–1324. https://doi.org/10.1016/S0266-3538(03)00103-9

Ratwani MM (2010) Composite materials and sandwich structures—a primer. Rto-En-Avt 156:1–16

Rohatgi PK (1993) Metal-matrix composites. Defence Sci J 43(4):323–349

Sands JM et al (2001) Environmental issues for polymer matrix composites and structural adhesives. Clean Prod Process 2(4):0228–0235. https://doi.org/10.1007/s100980000089

Satyanarayana KG, Arizaga GGC, Wypych F (2009) Biodegradable composites based on lignocellulosic fibers—an overview. Prog Polym Sci (Oxford) 34(9):982–1021. https://doi.org/10.1016/j.progpolymsci.2008.12.002

Shao Z, Vollrath F (2002) Surprising strength of silkworm silk. Nature 418(6899):741. https://doi.org/10.1038/418741a

Sharma R et al (2013) The impact of incinerators on human health and environment. Rev Environ Health 67–72. https://doi.org/10.1515/reveh-2012-0035

Smith GG, Barker RH (1995) Life cycle analysis of a polyester garment. Resour Conserv Recycl 14(3–4):233–249. https://doi.org/10.1016/0921-3449(95)00019-F

Stamboulis A et al (2000) Environmental durability of flax fibres and their composites based on polypropylene matrix. Appl Compos Mater 7(5–6):273–294. https://doi.org/10.1023/A:1026581922221

Surip SN et al (2016) Biodegradation properties of poly (lactic) acid reinforced by kenaf fibers. Acta Phys Pol A 835–837. https://doi.org/10.12693/aphyspola.129.835

Szczesna-Antczak M, Kazimierczak J, Antczak T (2012) Nanotechnology—methods of manufacturing cellulose nanofibres. Fibres Text Eastern Europe 91(2):8–12

Taya M, Arsenault RJ (2000) Metal matrix composites. Thermomechanical behavior. https://doi.org/10.1016/b978-0-08-031616-1.50001-7

Van de Weyenberg I et al (2006) Improving the properties of UD flax fibre reinforced composites by applying an alkaline fibre treatment. Compos A Appl Sci Manuf 37(9):1368–1376. https://doi.org/10.1016/j.compositesa.2005.08.016

Wollerdorfer M, Bader H (1998) Influence of natural fibres on the mechanical properties of biodegradable polymers. Ind Crops Prod 8(2):105–112. https://doi.org/10.1016/S0926-6690(97)10015-2

Woodford C (2017) Cermets. Available at https://www.explainthatstuff.com/cermets.html. Accessed 25 May 2018

Ye L et al (2017) Life cycle assessment of polyvinyl chloride production and its recyclability in China. J Clean Prod 142:2965–2972. https://doi.org/10.1016/j.jclepro.2016.10.171 Elsevier Ltd.

Yu PHF et al (1999) Transformation of industrial food wastes into polyhydroxyalkanoates. Water Sci Technol 365–370. https://doi.org/10.1016/s0273-1223(99)00402-3

Yu J (2001) Production of PHA from starchy wastewater via organic acids. J Biotechnol 86(2):105–112. https://doi.org/10.1016/S0168-1656(00)00405-3

Zini E et al (2004) Biodegradable polyesters reinforced with surface-modified vegetable fibers. Macromol Biosci 286–295. https://doi.org/10.1002/mabi.200300120

Zini E, Scandola M (2011) Green composites: an overview. Polym Compos. https://doi.org/10.1002/pc.21224

Zini E et al (2007) Bio-composite of bacterial poly(3-hydroxybutyrate-co-3-hydroxyhexanoate) reinforced with vegetable fibers. Compos Sci Technol 67(10):2085–2094. https://doi.org/10.1016/j.compscitech.2006.11.015

Zou Y, Xu H, Yang Y (2010) Lightweight polypropylene composites reinforced by long switchgrass stems. J Polym Environ 18(4):464–473. https://doi.org/10.1007/s10924-010-0165-4

Printed in the United States
By Bookmasters